Tumor Aneuploidy

Edited by Th. Büchner, C. D. Bloomfield,
W. Hiddemann, D. K. Hossfeld,
and J. Schumann

With Contributions by
M. Andreeff, B. Barlogie, D. B. von Bassewitz, R. Becher,
C. D. Bloomfield, Th. Büchner, F. Carbonell, B. Eagle,
T. M. Fliedner, A. Ganser, W. Göhde, E. Grundmann,
J. Hauss, H. Heimpel, G. Henze, W. Hiddemann,
D. Hoelzer, D. K. Hossfeld, U. Kaufmann, H. J. Kleinemeier,
V. Klinnert, H.-J. Langermann, M. R. Melamed, D. Miller,
K.-M. Müller, A. Redner, J. K. H. Rees, H. Riehm,
J. Ritter, A. Roessner, G. Schellong, J. Schumann,
P. Steinherz, S. Thongprasert, H.-J. Weh, B. Wörmann

With 49 Figures and 56 Tables

Springer-Verlag
Berlin Heidelberg GmbH

Professor Dr. Th. Büchner
Medizinische Klinik und Poliklinik der Universität
Albert-Schweitzer-Str. 33, D-4400 Münster, FRG

Professor Dr. Clara D. Bloomfield
Section of Medical Oncology
University of Minnesota School of Medicine
Coleman Leukemia Research and Treatment Center
Box 277 University of Minnesota Hospitals
Minneapolis, Minnesota 55455, USA

Priv.-Doz. Dr. W. Hiddemann
Medizinische Klinik und Poliklinik der Universität
Albert-Schweitzer-Str. 33, D-4400 Münster, FRG

Professor Dr. D. K. Hossfeld
Department of Oncology and Hematology
Medizinische Universitätsklinik Hamburg-Eppendorf
Martinistr. 52, D-2000 Hamburg 20, FRG

Dr. J. Schumann
Fachklinik Hornheide
Dorbaumstr. 300, D-4400 Münster, FRG

ISBN 978-3-540-15376-4   ISBN 978-3-642-70458-1 (eBook)
DOI 10.1007/978-3-642-70458-1

Library of Congress Cataloging in Publication Data
Main entry under title: Tumor aneuploidy. Includes bibliographies and index.
1. Leukemia — Genetic aspects. 2. Aneuploidy. I. Büchner, Th. II. Andreeff, Michael, 1943– . [DNLM: 1. Aneuploidy. 2. DNA, Neoplasms — analysis. 3. Leukemia — pathology. QZ 350 T925] RC643.T86 1985   616.99′419042   85-9897

© Springer-Verlag Berlin Heidelberg 1985

Typesetting: Graphischer Betrieb Konrad Triltsch, Würzburg
Offsetprinting: Brüder Hartmann, Berlin. Bookbinding: Lüderitz & Bauer, Berlin
2127/3020-543210

Preface

Chromosome abnormalities of cancer cells have been recognized for a long time, and have generally proven to be a highly specific marker of malignancy. The contributions collected in this book, "Tumor Aneuploidy", cover several major aspects of present knowledge concerning the occurrence and clinical significance of chromosome abnormalities as delineated by karyotype analyses or measurements of the cellular DNA content.

Certain non-random clonal chromosome losses, deletions and translocations appear to represent primary genetic lesions of malignancies and reflect their clonal origin. Secondary intraneoplastic genetic evolution is suggested by major clonal abnormalities of chromosome number and cellular DNA content. Both types of genetic changes have been reaching great relevance in cancer medicine, today.

Although the Philadelphia chromosome was first discovered in chronic myelocytic leukemia (CML), by Nowell and Hungerford in 1960, new banding techniques developed in the 1970's were needed to identify this abnormality as a translocation between chromosomes 9 and 22 (t(9; 22)). Soon thereafter, further non-random translocations were detected and attributed to special diseases like t(8; 21) and t(15; 17) to acute myeloid leukemias (AML) and t(9; 22), t(4; 11), t(8; 14) to acute lymphoblastic leukemia (ALL). Moreover, specific primary chromosome changes were found to be associated with special subclasses of acute leukemias such as t(8; 21) with differentiated myeloblast morphology, t(15; 17) with promyelocytic morphology and t(9; 11) with monocytic/monoblastic morphology in AML, and t(8; 14), t(2; 8), t(8; 22) in ALL with Burkitt type morphology, and isochromosome 17 q with CML blast crisis. In contrast to de novo leukemias, losses or deletions of chromosomes 5 or 7 were commonly found in secondary leukemias after radio- or chemotherapy of other malignancies. Interestingly, some chromosome breakpoints have been found to correspond with gene loci for immunoglobulins, T-cell receptors and sites of cellular oncogenes.

In contrast to structural aberrations, many numerical chromosome abnormalities and complex marker chromosomes in human malignancies probably reflect intraneoplastic evolution of aneuploid clones which may result from a genetic instability of the original tumor clone. These alterations are accompanied by corresponding clonal aberrations of the cellular DNA content (DNA aneuploidy). The incidence of DNA aneuploidy as detectable by modern flow cytometers has been shown to be related to the clinical and histologic grading in several diseases, with rare cases of DNA aneuploidy in low grade lymphomas, chronic leukemias, and basal cell skin carcinomas. The identification of DNA aneuploidy in some congenital melanocytic nevi, cases of ulcerative colitis and large-bowel polyps suggests a malignant transformation. High frequencies of 70 to over 90% DNA aneuploidies are reached in almost all carcinomas and sarcomas. The grade of DNA aneuploidy expressed by the DNA index also shows correlations with disease categories; acute leukemias and myelomas have near diploid DNA indices, testicular tumors cluster around the triploid value and other solid tumors distribute between near diploid and markedly hypertetraploid DNA indices. Multiple clone DNA aneuploidy is found in some tumors indicating intraneoplastic heterogeneity.

Both, the primary chromosome abnormalities and the DNA aneuploidies represent tumor cell markers which are highly specific for malignancy. Moreover, a specific karyotype or DNA abnormality may characterise an individual disease thus providing a reliable monitoring of therapeutic response, remission and relapse. Important relations have also been found between response or survival and specific karyotype and DNA abnormalities. Today, the highly sensitive detection of specific chromosome abnormalities, often based on only a few evaluable metaphases, is complemented by the high accuracy DNA flow cytometry of representative cell numbers.

The present volume contains reports which expand on the above concepts. Updated results and surveys are contributed by authors from 8 centers in Europe and in the United States. As a result of their cooperation this volume presents an overview of the current status of cytogenetics in human malignancies with particular emphasis on their clinical relevance.

The editors

Contents

Cytogenetic Studies in Acute Leukemia and in the Blast Phase of the Chronic Myelocytic Leukemia

F. Carbonell, V. Klinnert, A. Ganser, T. M. Fliedner, H. Heimpel, D. Hoelzer

Since 1960, when Nowell and Hungerford [1] observed for the first time the Philadelphia chromosome (Ph') in patients with chronic myelocytic leukemia (CML), the interest in cytogenetical studies in hematological disorders has increased. However, at the beginning of cytogenetical investigation no further abnormalities could be found with the commonly used techniques. In the early 1970s, a major innovation in cytogenetics occurred with the discovery of new techniques which allowed a better identification of chromosomes [2, 3]. With the new banding techniques it was possible, in addition to numerical alterations, to identify many structural abnormalities, particularly deletions and translocations in the leukemic karyotype. Recently, new culture methods have been developed which increase the number of leukemic metaphases available for analysis and at the same time allow a better morphological identification of the chromosome changes [4]. Recent progress indicates that the chromosome studies in leukemia could be useful for diagnosis, prognosis and therapy [5–7]. In our present report we want to discuss two aspects: the contribution of cytogenetical analysis to the diagnosis and classification of leukemic disorders and the influence of different types of cell cultures on the clonal expression of the leukemic cells.

Material and Methods

Patients

The patient population of this study consisted of 22 cases with acute lymphoblastic leukemia (ALL), 55 with acute myelogenous leukemia (AML) and 52 patients in the blast phase of the chronic myelocytic leukemia (CML). The diagnosis was made from case history, physical examination and cytochemical studies of bone marrow aspirates and peripheral blood smears.

Culture

Mononuclear cells were obtained from bone marrow, and peripheral blood samples by separation on a Percoll or Ficoll-Isopaque gradient. Suspension cultures were performed in McCoy's 5A medium supplemented with 20% fetal calf serum, with and without stimulation with 15% human placenta conditioned medium. Cultures

Abt. Klinische Physiologie und Arbeitsmedizin und Abt. Innere Medizin III der Universität Ulm, D-7900 Ulm/Donau, FRG

Tumor Aneuploidy
Büchner et al.
© Springer-Verlag: Berlin Heidelberg 1985

for myeloid (CFU-C) and erythroid (BFU-E) progenitor cells were performed using the methylcellulose technique [8].

Cytogenetics

From the unseparated bone marrow specimens chromosome preparations were performed by the direct technique. Suspension cultures were harvested for cytogenetical analysis after 2 days of incubation using standard methods [8].

Chromosome analysis of in vitro colonies was made after 10–14 days' growth from pooled colonies as described by Moore and Metcalf [9]. The preparations were stained by the G-banding technique [3]. Karyotypes were identified according to the International System for Human Cytogenetic Nomenclature [10]. The abnormality was considered clonal when at least two cells had a similar rearrangement or when 3 cells were lacking the same chromosome.

The Blast Phase of Chronic Myelocytic Leukemia

During the acute phase, about 80% of all patients developed additional chromosome aberrations superimposed on the karyotype with only the Ph' translocation (Table 1). The secondary aberrations were nonrandom, and at this stage of the disease trisomy 8, isochromosome 17q and a second Ph' chromosome are the most characteristic additional abnormalities observed. The results obtained in the present series are very similar to those reported by other authors [6, 7, 11, 12], showing that the changes of the karyotype are specific and have been demonstrated in a large number of patients.

The cytogenetic follow-up of patients with CML showed that the additional chromosome abnormalities could be observed in the progression of the leukemic disorder. At the initial phase, Ph' negative metaphases could be detected in the bone marrow and in a late phase it was possible to detect new atypical clones. The chromosomal changes characteristic for the acute phase could be found several months prior to any clinical evidence of the blastic crisis. In 31 patients with Ph' positive CML we have performed a follow-up of the karyotypic evolution during the chronic and/or the blastic phase. During the chronic phase, 7 out of 31 patients developed subpopulations of leukemic cells with additional chromosomal aberrations similar

Table 1. Summary of cytogenetic studies in patients with chronic myelocytic leukemia

Phase	Karyotype				
	Ph'/normal	Ph'	Ph' and +Ph', +8, i (17q)	Ph'-complex	Ph' and other aberrations
Chronic n = 130	17 (13%)	98 (75%)	4 (3%)	6 (5%)	5 (4%)
Acute n = 52	1 (2%)	7 (13%)	35 (68%)	2 (4%)	7 (13%)

2

to those observed in blast crisis. The change in the karyotype could be found two to nine months prior to any clinical evidence of blast crisis. However, in some reports [13, 14] the patients showed a change without subsequent development of blast crisis, reflecting probably a slow progression of the disease in some cases. The continued surveillance of the cytogenetic picture in CML may have an important bearing on the therapeutic approaches to be taken in this disease and, in particular, on the prediction of the onset of the blastic phase of CML and the treatment of this phase of the disease [15]. The correlation between chromosomal patterns, morphological type and prognosis of the blast phase has been studied by different authors [16–18] with very controversal results and requires further data and evaluation.

Acute Myeloblastic Leukemia (AML)

Abnormal karyotypes have been reported in approximately 50% of all patients with acute myelogenous leukemia [6, 7, 19, 22]. In our series we observed clonal chromosome aberrations in 25 out of 55 patients studied (45%) (Table 2). There were 8 patients with hypodiploid, 10 pseudodiploid and 7 with hyperdiploid cell clones. The most common abnormalities were monosomy 7 in the hypodiploid group and trisomy 8 in the hyperdiploid cases. However, the incidence of cytogenetic abnormalities will probably be significantly greater when techniques for culturing leukemic cells and for obtaining prophase chromosomes are generally applied [4, 23]. The chromosome banding studies in AML confirmed that many of the abnormalities, both numerical and structural, were of a nonrandom nature. Specific chromosome translocations were found which appeared to define subgroups of AML with different morphological and prognostic features. The two aberrations that appear to be closely associated with a block in a particular stage of leukemic cell maturation the t(8;21) associated with acute myelocytic leukemia with maturation and the t(15;17) seen in the acute promyelocytic leukemia. The translocation between chromosomes 8 and 21 [t(8;21)] has been observed in patients with a diagnosis of M2 (acute myeloblastic leukemia with maturation) according to the FAB classification [24]. The translocation is often associated with other chromosome aberrations, most commonly with loss of one sex chromosome. This translocation has been associated

Table 2. Summary of cytogenetic studies in patients with acute leukemia

Karyotype	AML	ALL
Normal	30	5
Abnormal	25 (45%)	17 (77%)
Hypodiploid	8	1
Pseudodiploid	10	8
Hyperdiploid	7	8
Ph' chromosome	1	5
Total	55	22

Table 3. Cytogenetic studies in patients with secondary acute leukemia

Patient	First disease	Type of leukemia	Karyotype
1	Non-Hodgkin-lymphoma	AML	46, XY, 18p+
2	Non-Hodgkin-lymphoma	AML	45, XX, −7
3	Morbus Hodgkin	AML	43–44 Chr. Mult. Aberration
4	Aplastic anemia	AML	45, XY, −7

with a good prognosis, but the presence of additional chromosome abnormalities (loss of one sex chromosome) may confer a poorer prognosis [25, 26].

The rearrangement involving the chromosomes 15 and 17 [t(15;17)] has been associated with acute promyelocytic leukemia (APL). This translocation has not been described in any other type of leukemia. As a result of this specificity of the t(15;17) for APL, it must now be accepted that the acute promyelocytic leukemia is morphologically heterogeneous. A form without the large granules (microgranular variant) has been described [27–30].

Structural aberrations involving the long arm of chromosome 11 have been frequently observed in acute monocytic and acute myelomonocytic leukemia. The abnormalities occur most frequently in children and show a great variation in the breakpoints and in the other chromosomes involved in the rearrangements [31]. Recently, an association between deletion or inversion of chromosome 16 and increase of bone marrow eosinophils in acute myelogenous leukemia has been established [32, 33].

Chromosome analysis could be of great importance in the diagnosis of treatment-induced leukemia (secondary leukemia). This is commonly associated with a prephase of pancytopenia with bone marrow abnormalities. The differentiation between therapeutic suppression of the hemopoiesis and early acute leukemia is often very difficult. Clonal chromosome aberrations have been identified in these patients (more than 90%) and they are nonrandom. Deletions of the long arm or loss of chromosome numbers 5 and 7 are usually associated with secondary leukemias [34, 35]. In contrast, patients without chromosome abnormalities usually do not progress to leukemia [5]. In our series all patients with secondary leukemia presented chromosomal aberrations (Table 3), usually hypodiploid karyotypes. The monosomy 7 was frequently associated with prior exposure to carcinogens or mutagenic substances. Other authors have noted the complexity of chromosomal changes in secondary leukemias, indicating considerable instability of the karyotypic picture as well as the complexity of the karyotypes themselves [36]. No cases of secondary leukemia characterized by t(8;21), t(15;17) or Ph'-chromosome have been reported. This might indicate that the mechanisms of the "novo" and the secondary leukemia are different [36].

Acute Lymphoblastic Leukemia (ALL)

Multiple cytogenetical studies of acute lymphoblastic leukemia have been published but few of these have utilized banding techniques, presumably because of the fuzzy and ill-defined morphology of the leukemic chromosomes. The data of the Third International Workshop on Chromosomes in Leukemia [37, 38] showed clonal abnormalities in 66% of the 330 cases analyzed. Pseudodiploidy was present in 53% of the cases, hyperdiploidy in 37% and hypodiploidy in 6% of the patients. The most common translocations were t(9;22) in 39, t(4;11) in 18 and t(8;14) in 14 patients. The correlation between karyotype and survival indicates that the chromosome aberrations are an independent prognostic factor in ALL. Specific chromosome abnormalities are significantly correlated with response to treatment and survival in both childhood and adult leukemias. In particular, the presence of t(4;11), t(8;14), Ph'-chromosome or hypodiploidy indicated a poor prognosis. Similar results have been observed independently by different authors [39–45]. Our study in 22 ALL patients also showed a high incidence of aneuploidy (Table 2). Chromosome aberrations were observed in 77% of the patients. The karyotypes in many of these cases showed a very complex arrangement, but some nonrandom translocations have been observed. The Ph'-chromosome has been detected in 5 cases. The translocations between the long arms of chromosomes 8 and 14 or 8 and 22 has been observed in two patients with B-cell leukemia. A translocation involving the chromosomes 4 and 11 was observed in an adult with a T cell leukemia with very high leukocyte count and very short evolution.

The Ph'-chromosome has been described in 20 to 30% of adult and in 5 to 10% of childhood ALL. The occurrence of the Ph'-chromosome in AML is more unusual (2%) [7, 12, 20, 37, 46, 48]. In our series the Ph'-positive acute leukemia constitutes 2% of adult AML and 20% of adult ALL. From the clinical point of view the presence of a Ph'-chromosome has been described as an unfavourable prognostic factor. Patients with Ph'-positive ALL showed a poor response to chemotherapy and had shorter survivals than cases with Ph'-negative ALL [37, 46, 47, 49]. The clinical, morphological and immunological characteristics of these leukemias are very heterogeneous. Of our 5 cases two patients showed a double blast cell population with myeloid and lymphoid cell membrane markers (mixed AL), one case did not express any marker (null ALL), the fourth case was a T-cell leukemia and the last patient had a common-ALL. Cases with a pre-B or B-cell differentiation have also been reported [50, 51]. The Ph'-positive acute leukemias can manifest very different phenotypic characteristics and probably the cells involved in these leukemias represent a very primitive hemopoietic precursor cell with potential to express various cell differentiation patterns [52–54].

Cytogenetical studies in patients with Burkitt's lymphoma or leukemia have established three types of translocations, t(8;14), t(2;8) and t(8;22). These translocations have been observed in a large number of Burkitt's tumours, independent of the type, and in patients with B-cell ALL, indicating that they are probably different clinical manifestations of the same disease [38, 41, 55–60]. Chromosomal translocations may have an important role in the pathogenesis of B-cell neoplasia. The chromosomes 2, 14 and 22 contain the loci for different human immunoglobulins and the cellular oncogene myc is normally carried on chromosome 8 [61–64].

The translocation between the long arms of chromosomes 4 and 11 has been described in patients with ALL presenting with marked leukocytosis, anemia, splenomegaly and very poor prognosis. It has also to be noted that this translocation is very often observed in neonatal and early childhood ALL [65–67].

5

Detection of Leukemic Clones Using Different Culture Systems

Cytogenetic Studies Using Short-Term Liquid Culture

Results of cytogenetic analysis performed in leukemic patients using a direct preparation of the bone marrow, and for comparison a short-term liquid culture, have shown that the frequency of detectable aneuploid karyotypes can be higher in preparations made after culture [23, 68–70]. This phenomenon has been frequently observed in patients with acute leukemia and t(15;17) or t(8;21) [71], but similar results have also been observed in other types of acute and chronic leukemias and in myelodysplastic syndromes [68–70]. These observations indicate that new cytogenetical methods may be of interest for a more precise classification of leukemic patients into distinct clinical and prognostic subgroups [4, 5]. The mechanisms of enrichment of cytogenetically abnormal metaphases in culture has not yet been clearly elucidated and requires further experimentation [8, 68, 69, 71].

Differences Between Bone Marrow and Blood Cultures

Another point to be stressed is the material used to perform the cytogenetic analysis. Some studies in which different hemopoietic organs were cytogenetically analyzed have demonstrated that new abnormal clones can be detected in the spleen and/or lymph nodes, but not in the bone marrow [72–74]. These results suggest that the origin of the acute transformation in CML patients may be extramedullary. We have analyzed concurrently bone marrow and blood in patients with CML (Table 4). We detected an isochromosome 17q in the blood culture in two cases whereas the marrow cultured under the same conditions showed cells with the Ph' chromosome as the only aberration. Two other patients also showed abnormal clones in blood which were not present in the marrow. These results are consistent with the hypothesis of an extramedullary localization of the new clones and the observation may be

Table 4. Comparison between the cytogenetic results in four patients with CML using bone marrow and blood cell cultures

Patient no.	Clone	No. of metaphases in	
		Bone marrow cultures	Blood cultures
22	Ph'	37	92
	Ph', i (17q)	0	27
156	Ph'	40	36
	Ph', i (17q)	0	14
235	44–47 Chr. Ph'	0	11
	51–56 Chr. 2 Ph'	24	38
367	Ph', i (17q)	28	28
	Ph', i (17q), +8	35	44
	Ph', i (17q), +8, +19	0	8

Table 5. Cytogenetical findings in three CML patients studied with different stem cell culture systems

Patient no.	Material	Number of Ph′ + /Ph′-metaphases			
		2-day liquid culture	CFU-C	BFU-E	CFU-GEMM
740	Blood	30/0	16/0	21/0	8/0
744	Blood	30/0	13/2	17/4	5/1
751	Blood	30/0	8/0	5/0	0

explained if one considers the myeloid cells in blood as a "pool" of leukemic stem cells that have originated at various bone marrow sites and also in other hemopoietic organs.

Cytogenetics on Hemopoietic Colonies

The chromosome studies on hemopoietic colonies derived from bone marrow or blood cell cultures from leukemic patients offers the opportunity to study the involvement of the progenitor cells in the leukemic process and to study the different growth requirements of cell subpopulations [8, 75].

Several hemopoietic stem cell culture methods are employed to detect the presence of different cell clones in patients with CML. As reported in a previous study [8], the presence of Ph′ chromosome negative metaphases has been observed in patients whose bone marrow or mononuclear blood cells were cultured in vitro for myeloid (CFU-C) or erythroid (BFU-E) stem cells, although short term liquid culture from the same material showed only Ph′ chromosome positive mitoses. New data in three patients (Table 5) using also cultures for pluripotent progenitor cells (mixed colonies or CFU-GEMM) confirmed the previously reported data and support the concept of a pluripotent hematopoietic stem cell origin of the disease. The presence of a mosaic of Ph′-positive and Ph′-negative cells could only be detected in patients short time after diagnosis. As the disease progressed we could no longer detect the presence of Ph′-negative cells. These data suggest that a population of functional but suppressed Ph′ chromosome negative stem cells could still be present in many patients at the initial phase of CML and with the progression of the disease this stem cell population is not only suppressed but has really disappeared.

References

1. Nowell PC, Hungerford DA (1960) A minute chromosome in human chronic granulocytic leukemia. Science 132:1497
2. Casperson T, Zeck L, Johansson C, Modest EJ (1970) Identification of human chromosomes by DNA-binding fluroescent agents. Chromosoma 30:215–227
3. Seabright M (1971) A rapid banding technique for human chromosomes. Lancet II:971–972

4. Yunis JJ, Bloomfield CD, Ensrud K (1981) All patients with acute nonlymphocytic leukemia may have a chromosomal defect. N Engl J Med 305:135–139
5. Bloomfield CD, Arthur DC (1982) Evaluation of leukemic cell chromosomes as a guide to therapy. Blood Cells 8:501–518
6. Mitelman F, Levan G (1981) Clustering of aberrations to specific chromosomes in human neoplasms. IV. A survey of 1871 cases. Hereditas 95:79–139
7. Sandberg AA (1980) The chromosomes in human cancer and leukemia. Elsevier North Holland, New York
8. Carbonell F, Hoelzer D, Grilli G, Issaragrisil, Harriss EB, Fliedner TM (1983) Chronic myelocytic leukemia: cytogenetical studies on haemopoietic colonies and diffusion chamber cultures. Scand J Haematol 30:486–491
9. Moore MAS, Metcalf D (1973) Cytogenetic analysis of human acute and chronic myeloid leukemic cells cloned in agar culture. Int J Cancer 11:143–152
10. ISCN (1978) An international system for human cytogenetic nomenclature. Cytogenet Cell Genet 21:309–402
11. Carbonell F, Benitez J, Prieto F, Badia L, Sanchez Fayos J (1982) Chromosome banding patterns in patients with chronic myelocytic leukemia. Cancer Genet Cytogenet 7:287–297
12. Rowley JD (1980) Ph'-positive leukemia, including chronic myelogeneous leukemia. Clinics in Haematology 9:55–127
13. Sadamori N, Matsunaga M, Jao E, Nishino K, Tomanaga Y, Tagawa M, Kusano M, Ichimaru M (1980) Chromosomes in the chronic phase of chronic granulocytic leukemia. Cancer Genet Cytogenet 1:299–310
14. Sonta S, Sandberg AA (1978) Chromosomes and causation of human cancer and leukemia. XXIX. Further studies on karyotypic progression in CML. Cancer 41:153–163
15. Hagemeijer A, Stenfert Kroeze WF, Abels J (1980) Cytogenetic follow-up of patients with nonlymphocytic leukemia. I. Philadelphia chromosome-positive chronic myeloid leukemia. Cancer Genet Cytogenet 2:317–326
16. Alimena G, Dallapiccola B, Gastaldi R, Mandelli F, Brandt L, Mitelman F, Nilsson PG (1982) Chromosomal, morphological and clinical correlations in blastic crisis of chronic myeloid leukemia. A study of 69 cases. Scand J Haematol 28:103–117
17. Olah E, Kiss A, Jako J (1980) Chromosome abnormalities, clinical and morphological manifestations in metamorphosis of chronic myeloid leukemia. Int J Cancer 26:37–45
18. Prigogina EL, Fleischman EW, Volkova MA, Frenkel MA (1978) Chromosome abnormalities and clinical and morphologic manifestations of chronic myeloid leukemia. Hum Genet 41:143–156
19. Golomb HM, Vardiman JW, Rowley JD, Testa JR, Mintz V (1978) Correlation of clinical findings with quinacrin-banded chromosomes in 90 adults with acute nonlymphocytic leukemia. N Engl J Med 299:613–619
20. First International Workshop on Chromosomes in Leukemia (1977) Chromosomes in acute non-lymphocytic leukemia. Br J Haematol 39:311–316
21. Philip P, Jensen MK, Killmann SA, Drivsholm A, Hansen NE (1978) Chromosomal banding patterns in 88 cases of acute non-lymphocytic leukemia. Leuk Res 2:201–212
22. Testa JR, Rowley JD (1980) Chromosomal banding patterns in patients with acute non-lymphocytic leukemia. Cancer Genet Cytogenet 1:239–248
23. Carbonell F, Fliedner TM, Kratt E, Sauerwein K (1979) Crecimiento de las celulas leucemicas en cultivo: seleccion de clonas citogeneticamente anormales. Sangre 24:1057–1060
24. Bennett JM, Catovsky D, Daniel HT, Flandrin G, Galton DAG, Gralnick HR, Sultan C (1976) Proposals for the classification of the acute leukemias. Br J Haematol 33:451–458
25. Second International Workshop on Chromosomes in Leukemia (1979) Cytogenetic, morphologic and clinical correlations in acute nonlymphocytic leukemia with t(8q–,21q+). Cancer Genet Cytogenet 2:99–102
26. Trujillo JM, Cork A, Ahearn MJ, Yonness EL, McCredie KB (1979) Hematologic and cytologic characterization of 8/21 translocation of acute granulocytic leukemia. Blood 53:695–706
27. Rowley JD, Golomb HM, Vardiman J, Fukuhara S, Dongherty Ch, Potter D (1977) Further evidence for a non-random chromosomal abnormality in acute promyelocytic leukemia. Int J Cancer 20:869–872

28. Second International Workshop on Chromosomes in Leukemia (1979) Chromosomes in acute promyelocytic leukemia. Cancer Genet Cytogenet 2:103–107
29. Golomb HM, Testa JR, Vardiman JW, Butler AE, Rowley JD (1979) Cytogenetic and ultrastructural features of de novo acute promyelocytic leukemia. Cancer Genet Cytogenet 1:69–78
30. Golomb HM, Rowley JD, Vardiman JW, Testa JR, Butler A (1980) Microgranular acute promyelocytic leukemia: a distinct clinical, ultrastructural and cytogenetic entity. Blood 55:253–259
31. Berger R, Bernheim A, Sigaux F, Daniel M, Valensi F, Flandrin G (1982) Acute monocytic leukemia. Chromosome studies. Leuk Res 6:17–26
32. Arthur DC, Bloomfield CD (1983) Partial deletion of the long arm of chromosome 16 and bone marrow eosinophilia in acute nonlymphocytic leukemia: a new association. Blood 61:994–998
33. LeBeau MM, Larson RA, Bitter MA, Vardiman JW, Golomb HM, Rowley JD (1983) Association of an inversion of chromosome 16 with abnormal marrow eosinophils in acute myelomonocytic leukemia. N Engl J Med 309:630–636
34. Berger R, Bernheim A, Daniel MT, Valensi F, Flandrin G (1981) Karyotypes and cell phenotypes in acute leukemia following other diseases. Blood Cells 7:293–299
35. Rowley JD, Golomb HM, Vardiman J (1977) Nonrandom chromosomal abnormalities in acute nonlymphocytic leukemia in patients treated for Hodgkin disease and non-Hodgkin lymphomas. Blood 50:759–770
36. Sandberg AA, Abe S, Kowalczyk JR, Zedgenidze A, Takenchi J, Kakati S (1982) Chromosomes and causation of human cancer and leukemia. L. Cytogenetics of leukemias complicating other diseases. Cancer Genet Cytogenet 7:95–136
37. Third International Workshop on Chromosomes in Leukemia (1981) Chromosomal abnormalities in acute lymphoblastic leukemia: structural and numerical changes in 234 cases. Cancer Genet Cytogenet 4:101–110
38. Third International Workshop on Chromosomes in Leukemia (1981) Clinical significance of chromosomal abnormalities in acute lymphoblastic leukemia. Cancer Genet Cytogenet 4:111–137
39. Bloomfield CD, Lindquist LL, Arthur D, McKenna RW, LeBien TW, Peterson BA, Nesbit ME (1981) Chromosomal abnormalities in acute lymphoblastic leukemia. Cancer Res 41:4838–4843
40. Cimimo MC, Rowley JD, Kinnealey A, Variakojis D, Golomb HM (1979) Banding studies of chromosomal abnormalities in patients with acute lymphocytic leukemia. Cancer Res 39:227–238
41. Kaneko Y, Rowley JD, Variakojis D, Chilcote RR, Check I, Sakurai M (1982) Correlation of karyotype with clinical features in acute lymphoblastic leukemia. Cancer Res 42:2918–2928
42. Secker-Walker LM, Lawler SD, Hardisty RM (1978) Prognostic implications of chromosomal findings in acute lymphoblastic leukaemia at diagnosis. Br Med J 2:1529–1530
43. Secker-Walker LM, Swansbury GS, Hardisty RM, Sallan SE, Garson OM, Sakurai M, Lawler SD (1982) Cytogenetics of acute lymphoblastic leukemia in children as a factor in the prediction of long-term survival. Br J Haematol 52:389–399
44. Williams DL, Tsiatis A, Brodeur GM, Look AT, Melvin SL, Bowman WP, Kalwinsky DK, Rivera G, Dahl GV (1982) Prognostic importance of chromosome number in 136 untreated children with acute lymphoblastic leukemia. Blood 60:864–871
45. Whang-Peng J, Knutsen T, Ziegler J, Leventhal B (1976) Cytogenetic studies in acute lymphocytic leukemia: special emphasis in long-term survival. Med Pediatr Oncol 2:333–351
46. Bloomfield CD, Peterson LC, Yunis JJ, Brunning RD (1977) The Philadelphia chromosome (Ph') in adults presenting with acute leukemia: a comparison of Ph'+ and Ph'− patients. Br J Haematol 36:347–358
47. Chessells JM, Janossy G, Lawler SD, Secker-Walker LM (1979) The Ph' chromosome in childhood leukaemia. Br J Haematol 41:25–41
48. Gustavsson A, Mitelman F, Olsson I (1977) Acute myeloid leukemia with the Philadelphia chromosome. Scand J Haematol 19:449–452

49. Bloomfield CD, Brunning RD, Smith KA, Nesbit ME (1980) Prognostic significance of the Philadelphia chromosome in acute lymphocytic leukemia. Cancer Genet Cytogenet 1:229–238

50. Alimena G, de Rossi G, Gastaldi R, Guglielsni G, Mandelli F (1980) B cell markers in Ph' positive acute lymphoblastic leukemia. Nouv Rev Fr Hematol 22:275–280

51. Vogler LB, Wust WM, Vinson PC, Sarrif A, Braltaim MG, Coleman MS (1979) Philadelphia chromosome positive pre-B cell leukemia presenting a blastic myelogenous leukemia. Blood 54:1164–1170

52. Boggs DR (1974) Hematopoietic stem cell theory in relation to possible lymphoblastic conversion of chronic myeloid leukemia. Blood 44:449–453

53. Catovsky D (1979) Ph'-positive acute leukaemia and chronic granulocytic leukaemia: one or two diseases? Br J Haematol 42:493–498

54. Sandberg AA, Kohno S, Wake N, Minowacha J (1980) Chromosomes and causation of human cancer and leukemia. XLII. Ph'-positive ALL: An entity within myeloproliferative disorders? Cancer Genet Cytogenet 2:145–174

55. Berger R, Bernheim A, Bronet JC, Daniel MT, Flandrin G (1979) t(8;14) translocation in a Burkitt's type of lymphoblastic leukemia (L3). Br J Haematol 43:87–90

56. Berger R, Bernheim A (1982) Cytogenetic studies on Burkitt's lymphoma-leukemia. Cancer Genet Cytogenet 7:231–244

57. Berghe van den H, Parloir G, Gosseye S, Englebienne V, Cornju G, Sokal G (1979) Variant translocation in Burkitt's lymphoma. Cancer Genet Cytogenet 1:9–14

58. Kaiser-McLaw B, Epstein AL, Kaplan HS, Hecht F (1977) Chromosome 14 translocation in African and North America Burkitt's lymphoma. Int J Cancer 19:482–486

59. Mitelman F (1981) Marker chromosome 14q+ in human cancer and leukemia. Adv Cancer Res 34:141–170

60. Mandara Y, Manolov G, Kieler J, Levan A, Klein G (1979) Genesis of the 14q+ marker in Burkitt's lymphoma. Hereditas 90:5–10

61. Dalla-Ferera R, Bregni M, Erikson J, Patterson D, Gallo RC, Groce CM (1982) Human c-myc oncogene is located on the region of chromosome 8 that is translocated in Burkitt lymphoma cells. Proc Natl Acad Sci (Wash) 79:7824–7827

62. Erickson J, Martinis J, Groce CM (1981) Assignment of the genes for human immunoglobulin chains to chromosome 22. Nature 294:173–175

63. Kirsch IR, Morton CC, Nakahara K, Leder P (1982) Human immunoglobulin heavy chain gene map to a region of translocations in malignant B lymphocytes. Science 216:301–303

64. McBride OW, Hieter PA, Hollis GF, Swan D, Otez MC, Leder P (1982) Chromosomal location of human kappa and lambda immunoglobulin light chain constant region genes. J Exp Med 155:1480–1490

65. Arthur DC, Bloomfield CD, Lindquist LL, Nesbit ME (1982) Translocation 4;11 in acute lymphoblastic leukemia: clinical characteristics and prognostic significance. Blood 59:96–99

66. Berghe van den H, David G, Broeckaert-Van Orshoven A, Lonwagie A, Verwilghen R, Casteels-Van Daele M, Eggermont E, Eeckels R (1979) A new chromosome anomaly in acute lymphoblastic leukemia (ALL). Human Genet 46:173–180

67. Parkin JL, Arthur DC, Abramson CS, McKenna RW, Kersey JH, Heideman RL, Brunning RD (1982) Acute leukemia associated with the t(4;11) chromosome rearrangement: ultrastructural and immunologic characteristics. Blood 60:1321–1331

68. Carbonell F, Grilli G, Fliedner TM (1981) Cytogenetic evidence for a clonal selection of leukemic cells in culture. Leuk Res 5:395–398

69. Knuutila S, Vuopio P, Elonen E, Szimes M, Kovanen R, Borgström GH, de la Chapelle A (1981) Culture of bone marrow reveals more cells with chromosomal abnormalities than the direct method in patients with hematological disorders. Blood 58:369–375

70. Waghray M, Epnes C, Rowley JD, Martin P, Testa JR (1981) Methods of processing marrow samples may affect the frequency of detectable aneuploid cells. Am J Hematol 11:409–415

71. Berger R, Bernheim A, Daniel MT, Valensi F, Flandrin G (1983) Cytogenetical types of mitosis and chromosome abnormalities in acute leukemia. Leuk Res 9:221–236

72. Hossfeld DK (1975) Chronic myelocytic leukemia. Cytogenetic findings and their relations to pathogenesis and clinic. Ser Haematol 8:53–72
73. Stoll C, Oberling F, Flori F (1978) Chromosome analysis of spleen and/or lymph nodes of patients with chronic myeloid leukemia. Blood 52:829–838
74. Zazzaria A, Baccarani M, Barbien E, Tura S (1975) Differences in marrow and spleen cell karyotype in early chronic myeloid leukemia. Blood 52:829–838
75. Löwenberg B, Hagemeijer A, Swart K (1982) Karyotypically distinct subpopulations in acute leukemia with specific growth requirements. Blood 59:641–645

The Clinical Usefulness of Chromosome Abnormalities in Acute Leukemia

C. D. Bloomfield

Introduction

Considerable excitement has been generated recently by newer techniques for the cytogenetic analysis of acute leukemia. Older, less precise, methods of analysis suggested that approximately 50% of patients with either ANLL or ALL had karyotype abnormalities in the leukemic cell population. However, it was thought that the chromosomal abnormalities detected were random rather than specific changes. The application of banding techniques, primarily quinacrine fluorescence (Q) and Giemsa (G), has permitted more detailed characterization of structural rearrangements within chromosomes. This has meant that gross morphologic alterations in the shape of the chromosome did not have to be present for abnormalities to be detected. It is now possible to precisely identify individual chromosomes despite deletions and translocations. Moreover, for unknown reasons, culture techniques, with or without methotrexate synchronization, increase relative to direct techniques the yield of mitoses containing the clonal chromosome abnormality, at least in ANLL (Berger et al. 1980a; Carbonell et al. 1979; Knuutila et al. 1980, 1981; Waghray et al. 1981; Fitzgerald et al. 1982; Hagemeijer et al. 1979; Yunis et al. 1981). By applying banding and various culture as well as direct techniques, and by obtaining adequate samples pretreatment of bone marrow, nonrandom chromosome changes can now be identified in the majority of patients with acute leukemia.

To date, chromosome analysis has been of clinical usefulness in acute leukemia in two major ways – in diagnosis and in prognosis. Specific chromosome abnormalities have been used to establish a diagnosis of leukemia. A prime example of this is seen with treatment-associated ANLL, where deletions of part of the long arm or absence of all of chromosome numbers 5 and 7 in the marrow of a pancytopenic patient following therapy for another malignancy may establish the diagnosis of ANLL, even in patients with a low marrow myeloblast count. A second example of the utility of cytogenetics for diagnosis is the evaluation of accumulations of cells in extramedullary sites. In the central nervous system, for example, it is not always clear whether the cells present are secondary to infection or leukemia. When the leukemia has a clonal chromosome abnormality, karyotype analysis can be used to establish the presence of leukemic cells.

Section of Medical Oncology, University of Minnesota School of Medicine, Coleman Leukemia Research and Treatment Center, Box 277, University of Minnesota Hospitals, Minneapolis, Minnesota 55455, USA

Tumor Aneuploidy
Büchner et al.
© Springer-Verlag: Berlin Heidelberg 1985

In addition to establishing a diagnosis of leukemia, chromosome analysis is increasingly being used for prognosis. It is now clear that the pretreatment bone marrow karyotype can be used, in both ANLL and ALL, to separate patients into groups with differing responses to treatment and survival. As a result, some therapists now use the pretreatment karyotype to select treatment for individual patients. For example, patients with karyotypes known to be associated with a short survival (e.g., Philadelphia chromosome positive ALL) are being treated in first remission with bone marrow transplantation.

As an increasing number of specific chromosome abnormalities are found to be associated with specific laboratory and clinical features in acute leukemia, their clinical usefulness will undoubtedly increase. In this brief review we will indicate the currently recognized specific chromosome abnormalities in acute leukemia and then review the major studies which have demonstrated that karyotype is an *independent* prognostic factor in both ANLL and ALL.

Clonal Chromosome Abnormalities in Acute Leukemia

Acute Nonlymphoblastic Leukemia

Using primarily culture techniques, a clonal chromosome abnormality has been found in more than 70% of cases of ANLL by several groups (Bloomfield and Arthur 1982; Yunis et al. 1981; Testa et al. 1983). The most common non-random *specific* chromosome abnormalities reported in *de novo* ANLL are indicated in Table 1. Many of these appear to have distinctive morphologic features (i.e., t(8;21) and M2, t(15;17) and M3 or M3 variant, t(9;11) and M5, abnormalities in 16q22 and morphologically abnormal eosinophils, t(6;9) and marrow basophilia). Others have distinctive clinical features (e.g., t(15;17) and disseminated intravascular coagulation, monosomy 7 and a high incidence of infection).

In addition, many nonspecific clonal chromosome abnormalities are seen in both ANLL and ALL. These are often grouped by modal number (i.e., the predominant number of chromosomes in cells with a range of chromosome numbers). Some of these nonspecific karyotypic subgroups appear to also have distinctive clinical and hematologic features (The Fourth International Workshop on Chromosomes in Leukemia 1984).

Acute Lymphoblastic Leukemia

Cytogenetic analysis with banding techniques has demonstrated clonal chromosome abnormalities in more than two-thirds of cases of ALL. In the largest series reported to date, using primarily direct marrow analysis, abnormalities were found in 66% of 330 newly diagnosed patients (Third International Workshop on Chromosomes in Leukemia 1981, 1983). Smaller studies have shown clonal abnormalities in approximately 80% of patients (Kaneko et al. 1982; Bloomfield et al. 1981). The chromosomal abnormalities that have been observed in ALL generally differ from those seen in ANLL. The most common specific chromosomal abnormalities identified in ALL have been t(9;22), t(4;11), and t(8;14) (The Third International Workshop on Chromosomes in Leukemia 1981, 1983; Bloomfield et al. 1977, 1980; Sandberg et al. 1980; Van den Berghe et al. 1979; Arthur et al. 1982; Berger et al. 1979).

Table 1. Specific chromosomal alterations in acute non-lymphoblastic leukemia

Morphologic/ Clinical features	Chromosome aberration	Selected references[a]
M2	t(8;21)(q22;q22). Often with loss of X or Y	Hart et al. 1971 Rowley 1973 SIWCL 1980
M3	t(15;17)(q22;q12–21)	Engel et al. 1967 Golomb et al. 1976, 1980 Rowley et al. 1977a, Hurd et al. 1982
	17q12 → q21. Variant translocations and deletions not involving chromosome 15	Yamada et al. 1983
M4 with abnormal eosinophils	del(16)(q22) inv(16)(p13q22)	Arthur and Bloomfield 1983 Le Beau et al. 1983
M4–M5	t(9;11)(p21;q23) 11q13 → q25. Deletions and translocations	Berger et al. 1980b Hagemeijer et al. 1982
M5	t(11;19)(q23;p12 or q12)	Morse et al. 1979
With basophilia	t(6;9)(p21–23;q33–34) 6p21 → pter. Deletions	Pearson et al. 1985
With infection	–7	Ruutu et al. 1977 Borgström et al. 1980
Secondary leukemia	5q12 → q32. Interstitial deletions of varying lengths –5 del(7)(q22 → qter),–7	Van den Berghe et al. 1974 Rowley et al. 1977b Mitelman et al. 1981
Poor prognosis	t(9;22)(q34;q11)	Bloomfield et al. 1977, 1980

[a] See de la Chapelle and Berger 1984 for a more complete list

Significant differences in frequency of clonal chromosome abnormalities have been found among immunologic subtypes of ALL (The Third International Workshop on Chromosomes in Leukemia 1981, 1983; Williams et al. 1984). All cases of B-ALL have had abnormalities identified, compared with 68% of non-T, non-B ALL and only 45% of T-ALL. The most common specific chromosomal abnormality in non-T, non-B ALL has been the Philadelphia chromosome (Ph[1]), most commonly t(9;22)(q34;q11). The second most common recurring chromosome abnormality has been t(4;11)(q21;q23); other recurring translocations that have been noted in this group include t(11;14)(q23;q32) and t(1;19)(q23;p13). The latter translocation has been associated with a pre-B phenotype (Williams et al. 1984). Among the non-specific chromosome abnormalities, the only one usually restricted to non-T, non-B ALL is a modal number of more than 50. Correlations of chromosome abnormalities with detailed immunologic phenotyping by a broad panel of monoclonal antibodies have not yet been done. Obviously, it would be of considerable interest to

15

determine if the specific immunologic phenotypes now recognized in ALL are correlated with specific karyotype groups.

Recurring specific chromosome abnormalities have less frequently been reported in T-ALL. Most cases of T-ALL have had a modal number of 46. A specific translocation t(11;14)(p13;q13) has been reported in T-ALL in children (Williams et al. 1984). Rare cases with the Ph¹ or t(4;11) have been reported. B-ALL almost always demonstrates an 8q− or a 14q+. Almost all adequately evaluated cases of B-ALL of L-3 type have had a specific translocation involving chromosome number 8 at band q24 (Berger and Bernheim 1982). The most common translocation, probably occurring in approximately 90% of cases, has been t(8;14)(q24;q32) (Manolova et al. 1979; Zhang et al. 1982). Two variant translocations have also been identified, t(8;22)(q24;q11) and t(2;8)(p12;q24). These variant translocations are of considerable interest because they appear correlated with the expression by the tumor cells of specific immunoglobulin light chains (Lenoir et al. 1982).

The relation between specific chromosome abnormalities and immunologic phenotype, however, requires further clarification. A number of the recurring chromosome abnormalities in B-cell leukemia or lymphoma are in regions to which immunoglobulin genes have been localized. Most striking is the localization of the heavy chain gene γ4 constant region to 14q32 (Kirsch et al. 1982). Moreover, recent studies demonstrate that immunoglobulin genes are moved as part of translocations in Burkitt cell lines (de la Chapelle et al. 1983; Erickson et al. 1982). In non-Burkitt's B lymphomas, associations of abnormalities of chromosome 2 at p12 and chromosome 22 at q11 with expression of specific light chains have not been found (Berger et al. 1983; Bloomfield et al. 1983). This suggests that the association of Burkitt's leukemia/lymphoma and immunoglobulin expression may occur because the DNA sequence involved in the malignant transformation of these cells is located on the long arm of chromosome 8 at band q24. The oncogene c-myc has been localized to chromosome 8q24 and has been demonstrated to translocate to chromosome 14 in Burkitt cell lines with t(8;14) (Taub et al. 1982; Dalla-Favera et al. 1982).

Prognostic Significance of Chromosome Abnormalities in Acute Leukemia

Acute leukemia is now clearly curable in many patients. In ANLL, 10–20% of adults treated with chemotherapy alone are 5-year disease-free survivors; 30–50% of children may be long-term disease-free survivors following intensive chemotherapy with or without subsequent marrow transplantation. In ALL, more than 50% of children and approximately 25% of adults are long-term disease-free survivors. Consequently, increasing emphasis in acute leukemia is being placed on identifying prognostic (or risk) factors and using them to develop classification systems that will allow the separation at diagnosis of those patients likely to be long-term disease-free survivors with current therapeutic approaches from those patients for whom present treatment is inadequate and new therapies must be developed. A number of recent studies indicate that results from chromosome analysis have prognostic utility in both ANLL and ALL.

Acute Nonlymphoblastic Leukemia

In ANLL, several reports have suggested that the presence or absence of normal metaphases in the pretreatment bone marrow aspirate may have prognostic significance, both with regard to the achievement of complete remission (Rothman et al. 1981; Golomb et al. 1978) and/or survival (Golomb 1982; The First International Workshop on Chromosomes in Leukaemia 1978; Hossfeld et al. 1979). Based on pretreatment karyotypes most commonly obtained from direct preparations of bone marrow, cases have been divided into those with all normal metaphases (NN), and those with a mixture of normal and abnormal metaphases (AN) or all abnormal metaphases (AA). Response and survival have generally been best among patients with NN karyotypes and poorest among those with AA karyotypes. When analyzed according to the FAB classification, this has been true for cases designated M1 and M2, but not for those classified as M4 and M5 (The First International Workshop on Chromosomes in Leukaemia 1978; Golomb et al. 1978). Using current intensive induction chemotherapy, drug resistance has been reported to be less frequent in patients with NN disease (Conjalka et al. 1981; Rothman et al. 1981). Moreover, it has been suggested that AA patients often fail induction therapy because of inadequate marrow regeneration once hypoplasia is induced (Rothman et al. 1981).

In contrast, the University of Minnesota experience suggests that the presence or not of a clonal chromosome abnormality per se does not predict response to chemotherapy or survival (Bloomfield and Arthur 1982). This is true whether all patients are considered or only those with M1 and M2 leukemias. Why these results differ from other studies with banding is unclear. However, chemotherapy used at the University of Minnesota was clearly more intensive than that used by some groups (Golomb et al. 1978), and our complete remission rate of 75% was higher than that for any of the earlier series. With more effective chemotherapy, the influence of prognostic factors often disappears. It is of interest that the two earlier studies that used relatively intensive treatment also reported no differences in duration of complete remission based on the presence or absence of clonal chromosome abnormalities (Rothman et al. 1981; Hossfeld et al. 1979).

We believe that the main reason that the University of Minnesota data differ from those reported by others is because the presence of clonal chromosome abnormalities per se is not a prognostic factor. Since most cases of ANLL probably have detectable abnormalities when studied with multiple cytogenetic techniques, the apparent prognostic significance of the presence of an abnormal clone is probably the prognostic significance of the cytogenetic technique used to detect the clone, not the presence or absence of the clone. The University of Minnesota group has used techniques that detect a higher frequency of abnormal clones than found by many groups (Bloomfield and Arthur 1982). Our techniques do not seem to result in a group of NN patients who respond better to treatment. However, even using the standard direct technique, certain abnormalities, such as the t(8;21), have been associated with a good prognosis. Our hypothesis is that eventually specific chromosome abnormalities will be found to have prognostic significance in ANLL, as appears to be the case in ALL. Some karyotypic subtypes will respond well to a given treatment regimen and others will not. Prospective studies of large numbers of patients will be required to test this hypothesis.

The data from the Fourth International Workshop appear to support the hypothesis of the clinical utility of subdividing ANLL into specific chromosome groups. Karyotype has been found to be an *independent* prognostic factor. The Fourth International Workshop on Chromosomes in Leukemia (1984) prospectively studied 660 newly diagnosed patients with *de novo* ANLL. The patients were grouped into 12 categories based first on specific chromosome abnormalities [i.e, t(8;21),t(15;17), −5 or 5q−, −7 or 7q−, similar abnormalities involving both 5 and 7, 11q abnormalities, +8, +21] and in the remaining cases with clonal chromosome abnormalities on the modal number of the abnormal clone (hyperdiploid, pseudodiploid, hypodiploid). This resulted in groups of patients who had significantly different presenting clinical and hematologic features, including FAB type, leukocyte count, percent peripheral myeloblasts, platelet count and disseminated intravascular coagulation.

Among 305 patients treated with Daunorubicin and Cytarabine based intensive combination chemotherapy, the above chromosome classification predicted frequency of complete remission (Table 2). Response rates of <50% were seen in patients who were hyperdiploid, had abnormalities in 5 and 7, had −5 or 5q− and were pseudodiploid. Response rates of >65% were seen in the t(15;17), +21, hypodiploid and t(8;21) groups. Follow-up is still limited since 48% of the patients were still alive when last analyzed. However, duration of first remission (p=0.04) and survival (p=0.005) were also predicted by the chromosome classification used by the Fourth Workshop. Remission duration and survival were short in patients with abnormalities in 5, 5 and 7, and 11q. The longest median survivals were seen in the hypodiploid, t(15;17) and −7 or 7q− groups. The presence or absence of abnormal metaphases (NN, AN, AA) did not correlate with response to initial induction therapy, but did correlate with first remission duration and survival. Patients with NN and AN leukemias had longer first remissions than those with AA

Table 2. Response to initial chemotherapy according to chromosome classification in 305 intensively treated patients with ANLL (Data from the Fourth International Workshop on Chromosomes in Leukemia, 1984)

Karyotype group	% CR[a]
Hyperdiploid	18
−5 or 5q− and −7 or 7q−	20
−5 or 5q−	42
Pseudodiploid	45
+8	50
Abnormal 11q	53
−7 or 7q−	57
Normal	60
t(15;17)	69
+21	71
Hypodiploid	71
t(8;21)	84
	p=0.01

[a] CR=complete remission

leukemias. Patients with NN leukemias survived longer than the other patients. However, when other major risk factors were considered for predicting survival, only the specific karyotype classification (not the NN, AN, AA classification) was an independent prognostic factor, predicting survival even when age, initial platelet count, presenting percent blood blasts and FAB type were considered.

Although the Fourth Workshop data are provocative, they require confirmation in a group of patients prospectively treated in a uniform fashion. Of the 660 patients with *de novo* leukemia in the Fourth Workshop, only 305 received intensive treatment including an anthracycline and cytarabine. Moreover, many different induction and post induction regimens were used. The only way to definitively assess the role of chromosome analysis using currently available techniques is to study a large group of uniformly treated patients prospectively.

Acute Lymphoblastic Leukemia

Several different features of cytogenetic analysis of the leukemic cell have been found to have prognostic significance in ALL including
1. the presence of specific chromosome abnormalities;
2. the chromosome number of the predominant abnormal clone;
3. the presence of translocations; and
4. the presence or absence of abnormal metaphases.

Certain specific chromosome abnormalities seem clearly to be associated with a poor prognosis. We first reported that ALL with the t(9;22) had a poor prognosis in 1975 and multiple studies have now confirmed the short survival of such patients (The Third International Workshop on Chromosomes in Leukemia 1981, 1983; Bloomfield et al. 1977, 1980; Sandberg et al. 1980). More recently both the t(4;11) and the t(8;14) have been found to predict poor response to treatment and short survival in ALL (The Third International Workshop on Chromosomes in Leukemia 1981, 1983; Arthur et al. 1982; Berger et al. 1979).

The chromosome number of the predominant abnormal clone was first suggested to have prognostic significance in ALL by Secker-Walker et al. (1978). They found that patients whose predominant abnormal clone was hyperdiploid (greater than 46 chromosomes) had the longest duration of remission. Others have confirmed that ALL with a modal number greater than 50 may constitute as much as 30% of childhood non-T, non-B ALL and appears to carry an unusually good prognosis (The Third International Workshop on Chromosomes in Leukemia 1981, 1983; Williams et al. 1982). Secker-Walker et al. (1978) also found that patients whose predominant abnormal clone was pseudodiploid (46 chromosomes) had significantly shorter durations of first remission than other patients. Others have confirmed the short remission durations seen in patients with pseudodiploid leukemias (The Third International Workshop on Chromosomes in Leukemia 1981, 1983; Bloomfield et al. 1981; Williams et al. 1982).

The fact that patients with a t(9;22), t(4;11) and t(8;14) did poorly suggested that translocations, in general, in ALL conferred a poor prognosis. This has been found to be the case (Bloomfield et al. 1981). However, data from The Third International Workshop on Chromosomes in Leukemia, suggest that it is only among children that patients with translocations survive a shorter period than patients with chromo-

19

some abnormalities other than translocations. Eventually, translocations which confer a favorable prognosis may be found in ALL, such as the t(8;21) in ANLL, but to date such have not been identified.

Finally, the adverse prognostic significance of the presence of abnormal metaphases in ALL has been suggested by some, but refuted by others (Bloomfield and Arthur 1982). The Third International Workshop on Chromosomes in Leukemia found a survival advantage for patients with all normal metaphases only among adults. The presence or absence of abnormal metaphases was not, however, found to be an independent prognostic variable when age, leukocyte count and FAB classification were considered.

The Third International Workshop on Chromosomes in Leukemia has demonstrated that subdividing the leukemic cell karyotype into ten groups based first on the presence or absence of specific chromosome abnormalities [i.e., t(9;22), t(4;11), t(8;14), other 14q+, 6q−] and in the remaining cases with clonal chromosome abnormalities on modal number of the abnormal clone (less than 46, 46, 47–50, greater than 50) has clinical utility. Achievement of complete remission, duration of first complete remission, and survival have differed significantly among these chromosome groups (Table 3). Multivariate analyses have demonstrated that karyotypic pattern is an independent prognostic factor for both duration of first complete remission and survival, even when age, initial leukocyte count, FAB type and broad immunologic phenotype are considered. Karyotype grouped in this way seems to confer more prognostic information than modal number, the presence of translocations per se or the presence or absence of abnormal metaphases.

The Third International Workshop study is the largest study evaluating banded chromosome analysis in patients with ALL. Although chromosome analysis ap-

Table 3. Response to initial chemotherapy and survival according to chromosome classification in 329 patients with ALL. (Data from the Third International Workshop on Chromosomes in Leukemia, 1981)

Karyotype group	% CR	Median Months[a]	
		CR	Survival
> 50	87	−[b]	34
6q−	75	21	29
Normal	93	19	24
46 abnormal	71	13	21
47–50	77	13	15
< 46	71	13	14
Ph[1] positive	53	8	13
14q+	53	24	9
t(4;11)	67	4	7
t(8;14)	60	4	5
p value	< 0.0001	0.002	0.001

[a] Estimated by life table analysis; [b] Curve has not yet fallen below the 50% mark and thus median cannot be estimated.

peared clinically useful, the study suffers from being retrospective. There is a consequent lack of homogeneity in treatment; seventeen different institutions using different treatment regimens were involved. Thus, it will be very important to determine if chromosome analysis is still clinically important in a group of prospectively studied patients with ALL treated with the best currently available therapy.

Conclusions

With recent improvements in cytogenetic methodologies, the number of specific chromosome abnormalities recognized in ANLL and ALL is rapidly increasing. Many of the recently recognized specific chromosome abnormalities seem to identify groups of patients that are quite homogeneous relative to many clinical and laboratory features. Thus, it is quite possible that in the future, karyotypes may be the major way in which we will diagnose and classify acute leukemia.

Although further study is required, the data indicate that karyotype as defined by current methodology is an *independent* prognostic factor for achievement of complete remission, duration of first complete remission and survival in both ANLL and ALL. Whether or not new approaches to treating patients in poor prognosis cytogenetic groups will improve their chances for long-term disease-free survival or cure needs to be explored. However, at the current time to provide patients with the most accurate information regarding prognosis, and to optimally evaluate clinical trials, banded cytogenetic analysis should be performed on all patients with ANLL and ALL at diagnosis.

References

Arthur DC, Bloomfield CD (1983) Partial deletion of the long arm of chromosome 16 and bone marrow eosinophilia in acute nonlymphocytic leukemia: A new association. Blood 61:994–998

Arthur DC, Bloomfield CD, Lindquist LL, Nesbit ME, Jr (1982) Translocation 4;11 in acute lymphoblastic leukemia: Clinical characteristics and prognostic significance. Blood 59.96–99

Berger R, Bernheim A (1982) Cytogenetic studies on Burkitt's lymphoma-leukemia. Cancer Genet Cytogenet 7:231–244

Berger R, Bernheim A, Brouet JC, Daniel MT, Flandrin G (1979) t(8;14) translocation in a Burkitt's type of lymphoblastic leukaemia (L3). Br J Haematol 43:87–90

Berger R, Bernheim A, Flandrin G (1980a) Hematologie. Absence d'anomalie chromosomique et leucémie aiguë: Relations avec les cellules medullaires normales. CR Acad Sci Paris 290:1557–1559

Berger R, Bernheim A, Weh HJ, Daniel MT, Flandrin G (1980b) Cytogenetic studies on acute monocytic leukemia. Leuk Res 4:119–127

Berger R, Bernheim A, Valensi F, Flandrin G (1983) 22q– and 8q– in a non-Burkitt lymphoma. Cancer Genet Cytogenet 8:91–92

Bloomfield CD, Arthur DC (1982) Evaluation of leukemic cell chromosomes as a guide to therapy. Blood Cells 8:501–518

Bloomfield CD, Peterson LC, Yunis JJ, Brunning RD (1977) The Philadelphia chromosome (Ph^1) in adults presenting with acute leukaemia: a comparison of Ph^1+ and Ph^1– patients. Br J Haematol 36:347–358

Bloomfield CD, Brunning RD, Smith KA, Nesbit ME (1980) Prognostic significance of the Philadelphia chromosome in acute lymphocytic leukemia. Cancer Genet Cytogenet 1:229–238

Bloomfield CD, Lindquist LL, Arthur D, McKenna RW, LeBien TW, Peterson BA, Nesbit ME (1981) Chromosomal abnormalities in acute lymphoblastic leukemia. Cancer Res 41:4838–4843

Bloomfield CD, Arthur DC, Frizzera G, Levine EG, Peterson BA, Gajl-Peczalska KJ (1983) Non-random chromosome abnormalities in lymphoma. Cancer Res 43:2975–2984

Borgström GH, Teerenhovi L, Vuopio P, de la Chapelle A, Van Den Berghe H, Brandt L, Golomb HM, Louwagie A, Mitelman F, Rowley JD, Sandberg AA (1980) Clinical implications of monosomy 7 in acute nonlymphocytic leukemia. Cancer Genet Cytogenet 2:115–126

Carbonell F, Fliedner TM, Kratt E, Sauerwein K (1979) Crecimiento de las células leucémicas en cultivo: Selección de clones citogenéticamente anormales. Sangre 24:1057–1060

Conjalka MS, Cuttner J, Wisniewski L, Goldberg JD, Elliott R, Reisman A, Desnick R, Holland JF, Berk PD (1981) Pretreatment marrow cytogenetic studies: a predictor of response to remission induction therapy in acute myelogenous leukemia. Proc Amer Soc Clin Oncol 22:487

Dalla-Favera R, Bregni M, Erikson J, Patterson D, Gallo RC, Croce CM (1982) Human c-myc onc gene is located on the region of chromosome 8 that is translocated in Burkitt lymphoma cells. Proc Natl Acad Sci 79:7824–7827

de la Chapelle A, Berger R (1984) Report of the committee on chromosome rearrangements in neoplasia and on fragile sites. Cytogenet Cell Genet 37:274–311

de la Chapelle A, Lenoir G, Boué J, Boué A, Gallano P, Huerre C, Szajnert M-F, Jeanpierre M, Lalouel J-M, Kaplan JC (1983) Lambda Ig constant region genes are translocated to chromosome 8 in Burkitt's lymphoma with t(8;22). Nucleic Acids Res 11:1133–1142

Engel E, McKee LC, Bunting KW (1967) Chromosomes 17–18 in leukaemias. Lancet I:42–43

Erickson J, Finan J, Nowell PC, Croce CM (1982) Translocation of immunoglobulin V_H genes in Burkitt lymphoma. Proc Natl Acad Sci USA 79:5611–5615

First International Workshop on Chromosomes in Leukaemia (1978) Chromosomes in acute non-lymphocytic leukaemia. Br J Haematol 39:311–316

Fitzgerald PH, Morris CM, Giles LM (1982) Direct versus cultured preparation of bone marrow cells from 22 patients with acute myeloid leukemia. Human Genet 60:281–283

Fourth International Workshop on Chromosomes in Leukemia (1984) Cancer Genet Cytogenet 11:332–350

Golomb HM (1982) Chromosome abnormalities in adult acute leukemias: biologic and therapeutic significance. In: Bloomfield CD (ed) Adult Leukemias 1. Martinus Nijhoff Publishers, Boston

Golomb HM, Rowley J, Vardiman J, Baron J, Locker G, Krasnow S (1976) Partial deletion of long arm of chromosome 17; a specific abnormality in acute promyelocytic leukemia? Arch Intern Med 136:825–826

Golomb HM, Vardiman JW, Rowley JD, Testa JR, Mintz U (1978) Correlation of clinical findings with quinacrine-banded chromosomes in 90 adults with acute nonlymphocytic leukemia. N Engl J Med 299:613–619

Golomb HM, Rowley JD, Vardiman JW, Testa JR, Butler A (1980) 'Microgranular' acute promyelocytic leukemia: a distinct clinical, ultrastructural, and cytogenetic entity. Blood 55:253–259

Hagemeijer A, Smit EME, Boatsma D (1979) Improved identification of chromosomes of leukemic cells in methotrexate-treated cultures. Cytogenet Cell Genet 23:208–212

Hagemeijer A, Hählen K, Sizoo W, Abels J (1982) Translocation (9;11)(p21;q23) in three cases of acute monoblastic leukemia. Cancer Genet Cytogenet 5:95–105

Hart JS, Trujillo JM, Freireich EJ, George SL, Frei E, III (1971) Cytogenetic studies and their clinical correlation in adults with acute leukemia. Ann Intern Med 75:353–360

Hossfeld DK, Faltermeier MT, Wendehorst E (1979) Beziehungen zwischen Chromosomenbefund und Prognose bei akuter nicht-lymphoblastischer Leukämie. Blut 38:377–382

Hurd DD, Vukelich M, Arthur DC, Lindquist LL, McKenna RW, Peterson BA, Bloomfield CD (1982) 15;17 translocation in acute promyelocytic leukemia. Cancer Genet Cytogenet 6:331–337

Kaneko Y, Rowley JD, Variakojis D, Chilcote RR, Check I, Sakurai M (1982) Correlation of karyotype with clinical features in acute lymphoblastic leukemia. Cancer Res 42:2918–2929

22

Kirsch IR, Morton CC, Nakahara K, Leder P (1982) Human immunoglobulin heavy chain genes map to a region of translocations in malignant B lymphocytes. Science 216:301–303

Knuutila S, Vuopio P, Borgström GH, de la Chapelle A (1980) Higher frequency of 5q-clone in bone marrow mitoses after culture than by a direct method. Scand J Haematol 25:358–362

Knuutila S, Vuopio P, Elonen E, Siimes M, Kovanen R, Borgström GH, de la Chapelle A (1981) Culture of bone marrow reveals more cells with chromosomal abnormalities than the direct method in patients with hematologic disorders. Blood 58:369–375

Le Beau MM, Larson RA, Bitter MA, Vardiman JW, Golomb HM, Rowley JD (1983) Association of inv(16)(p13q22) with abnormal marrow eosinophils in acute myelomonocytic leukemia: a unique cytogenetic-clinicopathologic association. N Engl J Med 309:630–636

Lenoir GM, Preud'homme JL, Bernheim A, Berger R (1982) Correlation between immunoglobulin light chain expression and variant translocation in Burkitt's lymphoma. Nature 298:474–476

Manolova Y, Manolov G, Kieler J, Levan A, Klein G (1979) Genesis of the 14q + marker in Burkitt's lymphoma. Hereditas 90:5–10

Mitelman F, Nilsson PG, Brandt L, Alimena G, Gastaldi R, Dallapiccola B (1981) Chromosome pattern, occupation, and clinical features in patients with acute nonlymphocytic leukemia. Cancer Genet Cytogenet 4:187–214

Morse H, Hays T, Peakman D, Rose B, Robinson A (1979) Acute nonlymphoblastic leukemia in childhood. Cancer 44:164–170

Pearson MG, Vardiman JW, LeBeau MM, Rowley JD, Schwartz S, Kerman SL, Cohen MM, Fleischman EW, Prigogina EL (1985) Increased numbers of marrow basophils may be associated with a t(6; 9) in ANLL. Am J Hematol 18:393–403

Rothman H, Preisler H, Reese P, Pothier L (1981) Karyotypic pattern influence on remission induction rate, duration of remission and survival in acute non-lymphocytic leukemia (ANLL). Proc Amer Soc Clin Oncol 22:489

Rowley JD (1973) Identification of a translocation with quinacrine fluorescence in a patient with acute leukemia. Ann Genet Paris 16:109–112

Rowley JD, Golomb HM, Dougherty C (1977a) 15/17 translocation, a consistent chromosomal change in acute promyelocytic leukaemia. Lancet I:549–550

Rowley JD, Golomb HM, Vardiman J (1977b) Non-random chromosomal abnormalities in acute non-lymphocytic leukemia in patients treated for Hodgkin's disease and non-Hodgkin lymphomas. Blood 50:759–770

Ruutu P, Ruutu T, Vuopio P, Kosunen T, de la Chapelle A (1977) Defective chemotaxis in monosomy-7. Nature 265:146–147

Sandberg AA, Kohno S, Wake N, Minowada J (1980) Chromosomes and causation of human cancer and leukemia. XLII. Ph^1-positive ALL: an entity within myeloproliferative disorders. Cancer Genet Cytogenet 2:145–174

Second International Workshop on Chromosomes in Leukemia (1980) Cytogenetic, morphologic, and clinical correlations in acute nonlymphocytic leukemia with t(8q–;21q+). Cancer Genet Cytogenet 2:99–102

Secker-Walker LM, Lawler SD, Hardisty RM (1978) Prognostic implications of chromosomal findings in acute lymphoblastic leukaemia at diagnosis. Br Med J 2:1529–1530

Taub R, Kirsch I, Morton C, Lenoir G, Swan D, Tronick S, Aaronson S, Leder P (1982) Translocation of the c-myc gene into the immunoglobulin heavy chain locus in human Burkitt's lymphoma and murine plasmacytoma cells. Proc Natl Acad Sci 79:7837–7841

Testa JR, Oguma N, Misawa S, Wiernik PH (1983) Chromosome abnormalities in acute leukemia. A higher incidence than previously assumed. Cancer Genet Cytogenet 9:305–306

Third International Workshop on Chromosomes in Leukemia 1980 (1981) Clinical significance of chromosomal abnormalities in acute lymphoblastic leukemia. Cancer Genet Cytogenet 4:111–137

The Third International Workshop on Chromosomes in Leukemia (1983) Chromosomal abnormalities and their clinical significance in acute lymphoblastic leukemia. Cancer Res 43:868–873

Van den Berghe H, Cassiman JJ, David G et al. (1974) Distinct haematological disorder with deletion of long arm of no. 5 chromosome. Nature 251:437

Van den Berghe H, David G, Broeckaert-Van Orshoven A, Louwagie A, Verwilghen R, Casteels-Van Daele M, Eggermont E, Eeckels R (1979) A new chromosome anomaly in acute lymphoblastic leukemia (ALL). Hum Genet 46:173–180

Waghray M, Eques C, Rowley JD, Martin P, Testa JR (1981) Methods of processing marrow samples may affect the frequency of detectable aneuploid cells. Am J Hematol 11:409–415

Williams DL, Tsiatis A, Brodeu GM, Look AT, Melvin SL, Bowman WP, Kalwinsky DK, Rivera G, Dahl GV (1982) Prognostic importance of chromosome number in 136 untreated children with acute lymphoblastic leukemia. Blood 60:864–871

Williams DL, Look T, Melvin SL, Roberson PK, Dahl G, Flake T, Stass S (1984) New chromosomal translocations correlate with specific immunophenotypes of childhood acute lymphoblastic leukemia. Cell 36:101–109

Yamada K, Sugimoto E, Amano M, Imamura Y, Kobota T, Matsumoto M (1983) Two cases of acute promyelocytic leukemia with variant translocations: the importance of chromosome no. 17 abnormality. Cancer Genet Cytogenet 10:93–99

Yunis JJ, Bloomfield CD, Ensrud K (1981) All patients with acute nonlymphocytic leukemia may have a chromosomal defect. N Engl J Med 305:135–166

Zhang S, Zech L, Klein G (1982) High-resolution analysis of chromosome markers in Burkitt lymphoma cell lines. Int J Cancer 29:153–157

Aneuploidy in Acute Myeloid Leukemia

J. K. H. Rees

In 1978 the British Medical Research Council (BMRC) opened the 8th trial for the treatment of acute myeloid leukemia. As part of the initial studies collaborators in 70 centres were asked to provide samples of blood or, when possible, bone marrow for cytogenetic analysis using Giemsa banding techniques. The analyses were performed at 8 cytogenetic laboratories throughout the British Isles, but the majority were carried out in the Department of Haematological Medicine in Cambridge.

During the period of 5 years up to the end of May 1983, when the study closed, over 500 samples have been examined in this way. The majority (over 80%) of the studies have been carried out on peripheral blood samples and there has been a relatively high rate of failure in obtaining analysable preparations. This is caused partly by the fact that the majority of samples have not been transported in culture medium and delays in delivery have also resulted in a high death rate in the leukemia cell population.

260 samples of blood or bone marrow from patients entered into the trial have been successfully examined using a Giemsa banding technique. All patients received the same remission induction therapy in this study which was designed for use in patients aged above 14 years (the median age of all patients entered into the study was 52 years). The remission induction therapy consisted of Daunorubicin, Cytosine Arabinoside and 6 Thioguanine in a "1 + 5" combination.

Criteria for the presence of a clone were the identification of at least 2 hyperdiploid or at least 3 hypodiploid cells with identical karyotypes.

The distribution of normal and abnormal karyotypes among this group of 260 patients is shown in Table 1.

The percentage of abnormal clones is similar to that which has been described in other series, but it is possible that this is an underestimate because studies were performed on blood samples in a high percentage of cases. The allocation to one of

Table 1. Cytogenetic analysis in acute myeloid leukemia

Analysis by karyotype group
Total no. of cases: 260

	No. of cases	%
All normal metaphases (NN)	144	55
Abnormal and normal metaphases (AN)	69	27
All abnormal metaphases (AA)	47	18

Department of Haematological Medicine, University of Cambridge

Tumor Aneuploidy
Büchner et al.
© Springer-Verlag: Berlin Heidelberg 1985

Table 2. Cytogenetic analysis by cytological subgroups in AML

	No. of cases	% of cases	% with chromosome abnormalities (AA + AN)
AMyl (M1 + M2)	141	54.2	45
APL (M3)	13	5	69
AMML (M4)	82	32	38
AMoL (M5)	15	6	53
EL (M6)	4	1.5	50
AMegL	5	2	60
Total	260		Av. 45

Table 3. Non-random numerical abnormalities in 116 AA + AN cases of AML

	No. of cases	% of abnormal cases
Trisomy 8	27	23%
Monosomy 7	9	8%
Monosomy X	8	7%
Missing Y	12	10%

Table 4. Remission rates and median survival by karyotype group in AML

	NN	AN	AA
No. of cases	138	66	44
Remission rate	73%	77%	61%
Median survival (months)	20	14	12

three categories – NN (all normal); AN (mixed normal and abnormal) and AA (all abnormal) – follows the system introduced by Sandberg (1980). A study of the cytogenetic features analysed by morphological type is of great interest as specific translocations have been found characteristic of certain subgroups of this form of leukemia. Table 2 shows that some F.A.B. subtypes carry a high probability of possessing an identifiable chromosome abnormality. F.A.B. group M3 is underrepresented among the total number in the series because patients with APL were not admitted to the trial for the first 2 years.

Numerical abnormalities are unequally represented among the 116 patients carrying a partially or totally abnormal karyotype. Table 3 shows that Trisomy 8 is the most common; the majority of cases demonstrating monosomy X or missing Y were associated with the 8;21 translocation, although there were 2 patients who had isolated missing Y abnormalities.

Remission rates and median durations of survival are not seriously affected by the presence or absence of abnormal clones when all cases are grouped together in the Sandberg forms (Table 4). Patients with AA characteristics had a remission rate of

Table 5. Non-random structural abnormalities in 116 AA + AN cases of AML

	Cytological subgroup	No. of cases	% of abnormal cases
t (8;21) + variants	M2 (and M1)	26	22%
t (15;17)	M3	8	7%
t (9;11)	M4, M5	4	3%

Table 6. Remission rates in patients with specific changes in AML

	No. of cases	Remission rate
t (8;21)	25	88%
t (15;17)	7	43%
Trisomy 8	25	60%
Monosomy 7 or 7q−	13	62%

Table 7. Features of specific translocations in AML

1. Higher incidence in younger patients (< 60 years old)
2. Characteristic cytological appearances
3. Characteristic clinical features
4. Break points related to sites of cellular oncogenes

61%; although lower than the 73% complete remission rate among those assigned the NN classification, the difference is not significant. A similar trend exists for survival, but this is also not statistically significant.

The most common non-random structural abnormalities found among the 116 patients was the 8;21 translocation first described by Rowley (1973). The majority of patients having this translocation were younger than average and satisfied the criteria for the diagnosis of AML F.A.B. type 2: 3 were allocated to F.A.B. type M1 (Table 5).

Although patients with t(8;21) translocations had a high remission rate (Table 6) the duration of remission was rather shorter than average (Swirsky et al. 1983).

This translocation is associated with highly characteristic morphological features and has a high incidence of extra-medullary disease (Sakurai et al. 1976; Swirsky et al. 1983). Only 7 patients with t(15;17) had been followed sufficiently long enough for an assessment of clinical response to be made. The condition is frequently associated with severe intravascular coagulation and death following severe haemorrhage was responsible for unsuccessful treatment in this small group. General features associated with specific translocations in AML are shown in Table 7.

Paediatric study. 33 children under 14 years with AML who were being treated in a similar way formed the basis of a further study. Unsuccessful attempts to obtain Giemsa-banded preparations were made in an additional 13 children out of a total of 83 cases referred for treatment over a period of 3 years.

Table 8. Cytogenetic abnormalities in children with AML

Karyotype abnormality	No.	% of abnormal cases
t (8;21)	9	43
+8	5	23
− 7 or 7q −	2	−
t (15;17)	1	−
Pseudo diploid	3	
Hyperdiploid	1	
Hypodiploid	1	

11 children had entirely normal karyotypes. 22 children had clonal abnormalities, the nature of which was listed in Table 8. The t(8;21) translocation was found in 9/22 abnormal cases and 4 of these had extra-medullary disease.

It seems increasingly apparent that the study of the cytogenetic characteristics of cases of acute myeloid leukaemia will produce useful prognostic information. It is important, however, as in the use of any prognostic factor, to define the type of treatment which is used in a particular group of patients as its significance may change as therapy changes.

The contributions which cytogenetic studies can make to the understanding of the pathogenesis of leukaemia and to the design of treatment programmes is rapidly being recognised. Cytogenetic studies should if possible be carried out in all cases of acute leukemia.

This study can be summarised as yielding the following information:
1. 45% of adult patients had detectable chromosome abnormalities; 66% were abnormal in a small children's study.
2. Exclusion of secondary AML and AML following pre-leukaemia resulted in a lower frequency of monosomy 5, 5q− and monosomy 7.
3. Remission rates were lower in AA cases than in AN or NN cases, but the differences were not significant.
4. Differences in survival between karyotypic groups were small − unless one considers individual karyotypes.
5. The role of cytogenetics as a prognostic factor in AML must take into account specific chromosome changes.

Acknowledgement

I am greatly indebted to Dr. John Matthews Principal Cytogeneticist in the Dept. of Haematological Mediane Cambridge and all the collaborating cytogeneticists in this MRC study for the kindness in providing the results reported in this study.

References

Rowley JD (1973) Ann Genet (Paris) 16:109–112
Sandberg AA (1980) The Chromosomes in Human Cancer and Leukaemia; Elsevier
Sakurai M, Sandberg AA (1976) Cancer 38:762–769
Swirsky DM, Li YS, Matthew JG, Flemans RJ, Rees JKH, Hayhoe FGJ (1984) British J Haematol 56:199–213

Chromosome Findings in Secondary Acute Leukemias

D. K. Hossfeld, H.-J. Weh

In 1970 several papers were published in the international medical literature pointing to an increased incidence of leukemia particularly in patients who had received chemotherapy because of multiple myeloma. In the same year, while working in the Roswell Park Memorial Institute in Buffalo, USA, the senior author, had the opportunity to analyse the chromosomes of leukemic cells of four such patients. Chromosome banding techniques were not available at that time. Nevertheless a number of remarkable findings became apparent: all cases were revealed to have profound numerical and structural anomalies, 3 out of 4 cases were hypodiploid, loss of chromosomes of groups 4–5 and 17–18 was noted in 2 cases, double minutes were seen in 2 cases [1]. During the following years more than 250 additional cases have been investigated; our early observations were confirmed, but more importantly, the introduction of banding techniques increased their significance considerably.

In this paper we will briefly present 6 more cases. The main aim, however, will be a discussion of some major findings so far published in the literature.

Three of our patients (Tables 1 and 2) had a carcinoma as primary tumor, the other three had Morbus Hodgkin. Patient No. 1 developed a second cancer which, apart from surgery, was not followed by additional therapy. Patients 2 and 3 had received X-ray therapy only, all others chemotherapy in addition.

Five and 7 years following X-ray therapy (cases 2 and 3), and 1½ to 2 years after termination of chemotherapy (cases 1, 4, 5, 6) patients revealed symptoms indicating a myelodysplastic syndrome (MDS). Between 2 and 11 months later MDS had progressed into frank acute leukemia. The type of leukemia was M1 in 4 cases and M6 in 2 cases.

The main karyotypic findings were: strict pseudodiploidy in case 6 due to a terminal deletion in the long arms of chromosome No. 11 (q23); hypodiploidy in the other cases with chromosome numbers between 40 and 45 per metaphase and rather flat main modes; loss of a chromosome No. 5 (or a B-chromosome in unbanded cases) was detected in 4 cases, loss of a chromosome No. 7 (or a C-chromosome) also in 4 cases; marker chromosomes, the origin of which could not be identified, were seen in 5 cases; among them were double minutes in 4 cases. Between 15 and 50 (mean 30) metaphases had been analysed in each case; however, normal metaphases could be found in 2 cases only.

Combination chemotherapy for remission induction was given to 3 patients only. For various reasons no chemotherapy was attempted in the others. Remissions

---

Dept. Oncology and Hematology, Med. Univ. Klinik Hamburg-Eppendorf, Martinistr. 52, D-2000 Hamburg 20, FRG

Tumor Aneuploidy
Büchner et al.
© Springer-Verlag: Berlin Heidelberg 1985

Table 1.

Case	Diagnosis	Date Diagnosis	Stage	Treatment	Treatment Periods
1.	Ca mamma	6/69	$T_1N_XM_0$	10.800 rads	69, 72, 79
	Ca colon	9/76	Dukes B	CMP[a], AC[a]	76–78
2.	Ca bladder	7/75	B	Part. cystectomy 2.000 (?) rads	75
3.	Ca mamma	9/74	$T_1M_0N_0$	4.800 rads	74
4.	M. Hodgkin	6/76	IIB$_S$	7.000 rads COPP[a]	77, 80 / 77
5.	M. Hodgkin	3/78	IIIB$_E$	8.000 rads COPP[a], ABVD[a]	78, 79 / 78, 79
6.	M. Hodgkin	3/79	IIIA	13.600 rads COPP[a]	79, 80 / 79

[a] C = cyclophosphamide; M = methotrexate; P = prednisone; A = adriamycin; B = bleomycin; V = vinblastine; O = vincristine; D = dacarbazine; P = procarbazine

Table 2.

Case	Date MDS[a]	Date leukemia	Type leukemia	Karyotype leukemia	Treatment	Result	Survival
1.	8/80	1/81	M6	42–44, XX; AN −B, −C, −E, −G, +M, +dm	No	–	4 months
2.	9/80	5/81	M1	43–45, XY; AA −6, −7, 17q−, dm	TAD[b]	No	7 months
3.	1/82	3/82	M1	43–45, XX; AA −5, 7q−, −18, −21 +m1, +m2	TAD	No	1 month
4.	5/81	6/82	M1	40–44, XY; AA −5, 7q−, t (7;15), t (6;12), −16, −18, −21 +dm	No	–	2 months
5.	4/82	7/82	M6	42–44, XY; AN −B, −20, +r, +dm	No	–	1 month
6.	1/81	2/82	M1	46, YX, 11q−; AA	TAD	No	2 months

[a] MDS = myelodysplastic syndrome
[b] TAD = thioguanine, cytosine arabinoside, daunorubicin

could not be obtained. The median survival time after acute leukemia had been diagnosed was 2 months.

If one attempts to screen the by now voluminous literature on secondary acute leukemia, the following leading clinical features become apparent: among the primary malignancies Hodgkin and non-Hodgkin lymphomas predominate, followed by multiple myeloma, breast cancer, ovarian cancer and bladder cancer. The median age of patients with secondary leukemia is 60 years. It is remarkable that secondary leukemia is exceedingly rare in children and young adults. At least two facts may explain this peculiar age-related phenomenon, namely the fact that alkylating agents are used less frequently for the treatment of childhood neoplasms, and furthermore that cancer as such is an age-related disease. It may also be that children are more resistant to the mutagenic action of cytostatics than adults.

Morphologically all types of leukemia have been observed, M2 and M6, however, predominated. A considerable proportion of the secondary leukemias (about 10%) does not fit into the FAB-classification; they resemble an acute myelodysplastic syndrome with a percentage of blasts between 20–40 and severely disturbed erythro- and megakaryocytopoiesis. The existence of secondary (therapy-induced) chronic leukemias, myeloid or lymphoid, is very questionable since among other objections, none of the cases so far described demonstrated any chromosomal peculiarities.

Chromosomal analysis allows a more precise definition of secondary leukemias than clinical and morphological findings. The leading cytogenetic features are: structural and/or numerical anomalies in 80–100% of the cases; hypo- and pseudodiploidy in 70–80% of the cases; partial or total loss of chromosomes No. 5 and 7, but also of No. 3 and 17 in 60% of the cases; marker chromosomes, particularly double minutes; multiclonality and, related to this, karyotypic changes during the course of the disease.

In order to understand this astonishing phenomenon of such characteristic chromosomal changes in secondary acute leukemias, a number of questions remain to be elucidated. First of all the question comes up as to the karyotype of such leukemia-patients who had a primary malignancy in their history for which, however, they had received neither radio- nor chemotherapy. A literature survey indicates that under those circumstances the karyotype does not differ from patients with so-called de novo leukemia. This is also true for the rare patients with simultaneously developing neoplasias of the hematopoietic system, e.g. AML and multiple myeloma [2, 3].

Another question is whether a relationship exists between the kind of chromosomal anomalies in secondary leukemias and the type of the primary tumor as well as the treatment modality of the primary tumor. No such relationship is presently apparent although the literature contains some hints that patients with a carcinoma and/or radiotherapy in their history may have more often anomalies of chromosome No. 7 than of No. 5 [4].

At this juncture it appears appropriate to briefly mention that acute leukemias following chronic myeloid leukemia or polycythemia vera exhibit distinctively different karyotypes than the secondary leukemias under discussion. This may serve as an additional indication for a spontaneous and not an iatrogenic transformation of such hematopoietic disorders.

Anomalies involving chromosomes No. 5 and 7 are of central interest since they are so characteristic for secondary leukemias. The question must be raised how frequent such anomalies are in leukemic patients who did not have a neoplastic disease in their previous history. The 4th International Workshop on Chromosomes in Leukemia has paid attention to this problem [5]. It was found that the –5/5q– and/or the –7/7q– constellation occurred in 50 out of 660 patients (7.5%) with de novo leukemia and in 32 out of 56 patients (57%) with secondary leukemia. The Workshop reconfirmed that these anomalies are rare in childhood-AML. If they are found in a seemingly de novo-AML, a subsequent deeper analysis of the patient's history revealed quite often an exposure to potentially carcinogenic or mutagenic substances [6, 7].

To assess the impact of –5/5q– and –7/7q– in secondary leukemias we have to ask for their relationship to other hematopoietic disorders. A literature survey discloses that 5q– as the sole anomaly can be seen in only 15% of the patients with acute leukemia; at least 65% of the patients are rather characterized by a certain type of refractory anemia the most prominent feature of which are micromegakaryocytes [8]. The combination of 5q– with additional anomalies is almost invariably associated with acute leukemia. With regard to monosomy 7 and 7q– the situation is less clear; both anomalies may be markers of preleukemia as well as leukemia. The constellation monosomy 7 plus additional anomalies, however, almost invariably goes along with secondary leukemia.

It can thus be stated that 5q– and –7 can be encountered in patients with myelodysplastic syndromes who do not transform into acute leukemia and who had not been exposed to mutagenes or carcinogenes. It is of interest that, on the other hand, secondary leukemias are preceded almost invariably by myelodysplastic syndromes. Finally, 5q– or –7 as the only anomaly are very rare in secondary leukemia. Being faced with such a complex situation it is presently difficult to define the exact role of radio-chemotherapy in the origin of these typical chromosomal changes in secondary leukemia.

Longitudinal chromosome studies in patients at risk could help to shed some light on this. Along these lines a number of studies had been performed in the past; however, most of them are non-contributory for various reasons:

1. in the majority of studies PHA-stimulated peripheral blood cells were analysed and not bone marrow cells;

2. the studies did not cover a biologically relevant period of time which should be years;

3. a number of different cytogenetic methods were used which reflect different biological phenomena (sister chromatid exchange, chromosomal breakage, numerical and structural chromosome anomalies). Results of such studies, mostly derived from PHA-stimulated lymphocytes of ALL-children [9–12], some from lymphocytes of cancer-patients [13, 14] and some from bone marrow cells of cancer-patients [15, 16], are summarized in Table 3. They demonstrate that during chemotherapy bone marrow and blood cells will reveal an increased SCE-rate, an increased breakage rate as well as karyotypic anomalies. One to four years after the end of chemo-

Table 3.

		SCE	Breakage	Karyotype
Bone marrow	During therapy	+	+	+
	After therapy	–	–	–
Lymphocytes	During therapy	+	+	+
	After therapy	+	–	
			– / +	+

therapy bone marrow cells were normal, no matter what cytogenetic method had been applied while lymphocytes did continue to show increased breakage and stable karyotypic changes. Thus, as yet such studies do not provide a cytogenetic basis as to the pathomechanism of acute leukemias following radiochemotherapy of primary neoplasias.

Supported by the Hamburger Krebsgesellschaft. The authors thank Prof. Hausmann, St. Georg Hospital, Hamburg, for his permission to study his patients.

References

1. Hossfeld DK, Holland JF, Cooper RG, Ellison RR (1975) Chromosome studies in acute leukemias developing in patients with multiple myeloma. Cancer Res 35:2808–2813
2. Matsuzaki H, Yamaguchi K, Hara H, Mitsuya H, Kawano F, Araki K, Tanaka R, Kishimoto S (1983) Simultaneous occurrence of acute myelogenous leukaemia and multiple myeloma without previous chemotherapy. Scand J Haematol 30:278–286
3. Thiagarajan P (1979) Simultaneous presentation of multiple myeloma and acute leukemia in the absence of previous chemotherapy – Report of a case. M Sinai J Med (N.Y.) 46:360–363
4. Sandberg AA, Abe S, Kowalczyk JR, Zedgenidze A, Takeuchi J, Kakati S (1982) Chromosomes and causation of human cancer and leukemia: L. cytogenetics of leukemias complicating other diseases. Cancer Genet Cytogenet 7:95–136
5. The Fourth International Workshop on Chromosomes in Leukemia, 1982 (1984) Cancer Genet Cytogenet 11:249–360
6. Mitelman F, Nilsson PG, Brandt L, Alimena G, Gastaldi R, Dallapiccola B (1981) Chromosome pattern, occupation, and clinical features in patients with acute nonlymphocytic leukemia. Cancer Genet Cytogenet 4:197–214
7. Golomb HM, Alimena G, Rowley JD, Vardiman JW, Testa JR, Sovik C (1982) Correlation of occupation and karyotype in adults with acute nonlymphocytic leukemia. Blood 60:404–411
8. Van den Berghe H, Cassiman J-J, David G, Fryns J-P, Michaux J-L, Sokal G (1974) Distinct haematological disorder with deletion of long arm of No. 5 chromosome. Nature 251:437–438
9. Otter M, Palmer CG, Baehner RL (1979) Sister chromatid exchange in lymphocytes from patients with acute lymphoblastic leukemia. Human Genet 52:185–192
10. Miller RC, Hill RB, Nichols WW, Meadows AT (1978) Acute and long-term cytogenetic effects of childhood cancer chemotherapy and radiotherapy. Cancer Res 38:3241–3246
11. Robison LL, Arthur DC, Ball DW, Danzl TJ, Nesbit ME (1982) Cytogenetic studies of long-term survivors of childhood acute lymphoblastic leukemia. Cancer Res 42:4289–4292
12. Inoue S, Brown L, Ravindranath Y, Ottenbreit MJ (1982) Normal sister chromatid exchange frequency in long-term survivors with acute leukemia. Cancer Res 42:2906–2908
13. Gebhart E, Lösing J, Wopfner F (1980) Chromosome studies on lymphocytes of patients under cytostatic therapy. I. Conventional chromosome studies in cytostatic interval therapy. Human Genet 55:53–63
14. Gebhart E, Windolph B, Wopfner F (1980) Chromosome studies on lymphocytes of patients under cytostatic therapy. II. Studies using the BUDR-labelling technique in cytostatic interval therapy. Human Genet 56:157–167
15. Nowell P, Glick JH, Bucolo A, Finan J, Creech R (1981) Cytogenetic studies of bone marrow in breast cancer patients after adjuvant chemotherapy. Cancer 48:667–673
16. Patil SR, Corder MP, Jochimsen PR, Dick FR (1980) Bone marrow chromosome abnormalities in breast cancer patients following adjuvant chemotherapy. Cancer Res 40:4076–4080

Chromosomal Heterogeneity in Solid Tumors

R. Becher

Dedicated to Avery A. Sandberg and his wife Maryn

Introduction

The clonal identity of malignant cells can be described according to their genotype and phenotype (Fig. 1).

The *genotype* may be defined quantitatively by the DNA content of cells and qualitatively by the chromosomal constitution. These two methods, measurement of the DNA content and chromosome analysis, are complementary to each other. In the past the genotype of cancer cells was considered to be stable; however, recently the genotype of cancer cells has been frequently found to display qualitative as well as quantitative alterations due to changes in the amount of DNA content and structural chromosomal abnormalities. An alteration of the karyotype may first of all result from neoplastic transformation or even be its cause. Later on, alterations of the DNA content and/or new chromosomal abnormalities may appear as a result of clonal evolution.

Besides the genotypic characterization, a malignant clone can also be described by its *phenotype*. Although phenotypic definitions of neoplastic clones are easily available, it must be kept in mind that clones identical due to their genotype are able to express a great variety of different phenotypes, resulting from differential activation of single genes which may be directed by a hypothetical superior operating system. The large number of phenotypes which can actually emerge from the same genome is demontrated by the morphological and functional heterogeneity of human somatic cells, thus illustrating the shortcomings of a mere description of clonal identity according to the phenotype. At present, methods for phenotypic characterization include observations on cell shape and size, growth characteristics and surface markers. In general, physical, chemical and/or functional parameters are utilized to define clonal identity by phenotypic features. Thus, a precise clonal definition requires both a phenotypic and a genotypic characterization of cells.

The following parameters can be used for a cytogenetic definition of malignant cells:

1. Total number of chromosomes
2. Numerical chromosome aberrations
   (Additional or missing normal chromosomes identified by banding)

Innere Universitäts- und Poliklinik (Tumorforschung), Westdeutsches Tumorzentrum, Hufelandstr. 55, D-4300 Essen 1, FRG

Tumor Aneuploidy
Büchner et al.
© Springer-Verlag: Berlin Heidelberg 1985

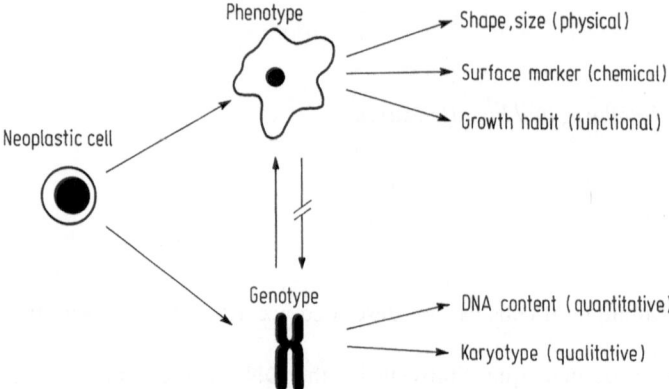

Fig. 1. Clonal identity as defined by phenotypic and genotypic features

3. Structural chromosome aberrations e.g. marker chromosomes
   (Abnormal chromosomes identified by an abnormal banding pattern and analysed according to their origin)

Total Number of Chromosomes in Tumor Cells

The majority of cytogenetic data published in the field of solid tumors refer to abnormal chromosome counts. The evaluation of the number of chromosomes per
metaphase is a rather simple approach requiring no detailed banding analysis.
Therefore, it offers the advantage of comparability of data published before banding
techniques became available.

The number of chromosomes in solid tumors has been shown to be usually in the
hyperdiploid range. During clonal evolution cells apparently tend to increase their
number of chromosomes as recently shown by us [1]. This kind of clonal evolution
may be associated with an accumulation of particular genes which lead to a growth
advantage of the cancer cells in their specific environment [2]. Thus, a general shift
from hyperdiploid to triploid or tetraploid modes can be observed in association
with tumor progression. Clonal chromosomal evolution may also be linked to a
newly acquired ability of tumors to metastasize; while this hypothesis appears
plausible, up to now no conclusive data are available to prove this concept.

The phenomenon of clonal evolution itself implicates clonal heterogeneity of cancer cells which are supposed to be subject to permanent competition among emer-
ging subpopulations.

Chromosome Abnormalities

The number of chromosomes alone is a rather rough parameter for clonal identification. Only the analysis of structural abnormalities in neoplastic cells provides qualitative information.

Most frequently, abnormal marker chromosomes originate from one or several translocations, which implies that a lost part of one chromosome becomes part of another one. This mechanism does not necessarily result in any measurable change in the amount of nuclear DNA. This kind of qualitative DNA alteration cannot be detected by means of flow cytometry. A well known example for a balanced translocation is the Philadelphia (Ph^1) chromosome, positive CML, where part of the long arm of chromosome number 22 is translocated to the terminal end of the long arm of chromosome number 9, forming the so-called standard translocation.

More rarely, a deletion of part of a single chromosome can occur as found in the 5q-syndrome, clinically characterized by severe anemia and an excess of blast cells.

A third known mechanism of alteration consists of duplication of a specific area in a given chromosome or an accumulation of multiplied single genes forming homogeneously stained regions (HSR) during clonal evolution of neoplastic cells [2].

All the structural changes mentioned can be traced to specific chromosomes.

Clonal Origin of Abnormal Chromosomes

The clonal origin of abnormal chromosomes is proven by their non-random formation. In solid tumors, in contrast to leukemias, a high number of abnormal chromosomes can frequently be found. Since they are not equally distributed in all metaphases of a tumor, they represent clonal heterogeneity of the genotype. It may be assumed that besides those specific abnormalities which are consistently found in all cells, there may be additional ones which are products of clonal evolution and which may indicate the development of tumor sublines.

Recently, we studied directly disaggregated material of malignant melanomas and bladder cancer cell lines utilizing a high resolution technique [5]. The detailed

Table 1. Frequency of marker chromosomes in malignant melanoma (Case # 2) based on the analysis of 13 banded karyotypes. Six karyotypes contained all 4 marker chromosomes

M1	M2	M3	M4
✕	✕	✕	✕
✕	✕	✕	✕
✕	✕	✕	✕
✕	✕	✕	✕
✕	✕	✕	✕
✕	✕	✕	✕
✕	✕	✕	
✕	✕	✕	
✕	✕	✕	
✕	✕	✕	
✕	✕	✕	
✕	✕	✕	
✕	✕	✕	

Total of 13 karyotypes

analysis of the karyotypes and the description of the markers found has been published elsewhere [3, 4]. Since it was possible to identify all marker chromosomes, we have been able to establish a distribution pattern of markers per karyotype. While the melanoma case 2 (Table 1) was characterized by only two markers and one side line, all other cases showed a rather high number of markers with a distinct distribution pattern (Tables 2, 3 and 4). The common feature in all cases was a certain number of cells which contained all markers together indicating that all cells originated from the same stem line. The heterogeneity in the distribution pattern of marker chromosomes is compatible with the theory of clonal evolution in solid tumors and suggests that genetic stability probably does last only for a limited period of time. Even stable culture conditions as given in the bladder cancer cell line studied apparently cannot assure a stable monoclonal condition.

Therefore, based on our data and on the current literature, we recommend a detailed chromosome analysis for clonal identification of neoplastic cells with special regard to the detection of genotypic heterogeneity.

Table 2. Frequency of marker chromosomes in malignant melanoma (Case # 1). A total of 14 markers was found in 8 karyotypes analysed. Three metaphases contained all markers. The markers M2, 3 and 4 were found in all metaphases

M1	M2	M3	M4	M5	M6	M7	M8	M9	M10	M11	M12	M13	M14
×	×	×	×	×	×	×	×	×	×	×	×	×	×
×	×	×	×	×	×	×	×	×	×	×	×	×	×
×	×	×	×	×	×	×	×	×	×	×	×	×	×
×	×	×	×	×	×	×	×	×	×	×		×	
×	×	×	×	×	×	×	×	×	×	×		×	
	×	×	×	×			×	×	×			×	
	×	×	×				×	×					
	×	×	×										

Total of 8 karyotypes

Table 3. Frequency of marker chromosomes in the bladder cancer cell line T-24 as analysed in 8 banded karyotypes. Only one cell contained all markers, whereas markers 4, 5, 6, 8, 12 and 13 were detected in all karyotypes

M1	M2	M3	M4	M5	M6	M7	M8	M9	M10	M11	M12	M13
×	×	×	×	×	×	×	×	×	×	×	×	×
		×	×	×	×	×	×	×	×	×	×	×
		×	×	×	×		×	×	×	×	×	×
		×	×	×	×		×	×	×	×	×	×
		×	×	×	×		×	×	×	×	×	×
		×	×	×	×		×	×	×	×	×	×
		×	×	×	×		×	×			×	×
			×	×	×		×				×	×

Table 4. Frequency of markers in bladder cancer cell line MG-U1 (EJ) as found in 24 karyotypes. A total of 14 markers was seen. One cell contained all markers and markers 6, 7 and 12 were found in all metaphases

M1	M2	M3	M4	M5	M6	M7	M8	M9	M10	M11	M12	M13	M14
×	×	×	×	×	×	×	×	×	×	×	×	×	×
×		×	×	×	×	×	×	×	×	×	×	×	×
		×	×	×	×	×	×	×	×	×	×	×	×
		×	×	×	×	×	×	×	×	×	×	×	×
		×	×	×	×	×	×	×	×	×	×	×	×
		×	×	×	×	×	×	×		×	×	×	×
		×	×	×	×	×	×	×			×	×	×
		×	×	×	×	×	×	×			×	×	×
		×	×	×	×	×		×			×	×	×
		×	×	×	×	×		×			×	×	×
			×	×	×	×		×			×	×	×
			×	×	×	×		×			×	×	×
			×	×	×	×		×			×	×	×
			×	×	×	×		×			×	×	×
			×	×	×	×		×			×	×	×
			×	×	×	×		×			×	×	×
					×	×		×			×	×	×
					×	×					×		×
					×	×					×		×
					×	×					×		×
					×	×					×		×
					×	×					×		
					×	×					×		
					×	×					×		

References

1. Becher R, Sandberg AA (1983) Entwicklungen, Befunde und Perspektiven in der Zytogenetik solider Tumoren. In: Schmidt CG (Ed) Aktuelle Probleme der Hämatologie und internistischen Onkologie. Springer, Berlin Heidelberg New York
2. Becher R (1984) Chromosomale Befunde und Chemotherapieresistenz. In: Seeber S, Osieka R, Sack H, Schönenberger H (Eds) Das Resistenzproblem bei der Chemo- und Radiotherapie maligner Tumoren. Beiträge zur Onkologie, Bd 18, S 212–220. Karger, Basel
3. Becher R, Gibas Z, Karakousis C, Sandberg AA (1983) Nonrandom chromosome changes in malignant melanoma. Cancer Res 43:5010–5016
4. Becher R, Gibas Z, Lin JC, Lin CW, Prout GR, Pontes JE, Sandberg AA Non-random cytogenetic changes in six human bladder cancer cell lines. (Submitted)
5. Yunis JJ (1976) High resolution of human chromosomes. Science 191:1268–1270

References

DNA Aneuploidy – A Common Cell Marker in Human Malignancies and Its Correlation to Grade, Stage and Prognosis

Th. Büchner[1], W. Hiddemann, J. Schumann, W. Göhde, B. Wörmann, J. Ritter,
H. J. Kleinemeier, D. B. von Bassewitz, A. Roessner, K.-M. Müller, E. Grundmann

Introduction

Numerous cytogenetic [1, 14, 30] as well as single cell cytophotometric [31] studies
have revealed karyotype or cellular DNA content abnormalities as a typical finding
in various malignant diseases. Reliable detection of clonal DNA aneuploidy, however, was made possible only by modern flow cytometers (FCM) [9, 10] combining
high accuracy DNA fluorometry with a rapid flow through system, thus providing
representative DNA distribution of cell populations. On the basis of FCM measurements in various malignancies by several investigators (overview see [3]) DNA
aneuploidy may be regarded the most common specific cell marker in cancer and
leukemia. By means of FCM, DNA aneuploidies were first observed in acute
leukemias [6], and some were later described in a series of different solid tumors by
the MD Anderson Hospital group in Houston and by our group [2, 32].

In the present multiinstitutional study at the University of Münster, answers
should be given regarding DNA aneuploidy, its incidence, grade, multiclonality and
heterogeneity in correlation to tumor types, dignity, grade of malignancy, tumor
stage and patient's prognosis. For this purpose DNA histogram analyses were conducted by FCM in 2413 cases of 13 different malignancies and in 776 benign lesions
or samples at 3 clinical departments in cooperation with the Department of Pathology. In addition, data from bone marrow samples of multicenter studies on acute
leukemias are included.

Material and Methods

Single cell suspensions of solid tissue samples and bone marrow biopsies were obtained by mechanical mincing, pepsination and filtration through a nylon mash
gaze. Samples from effusions, bone marrow aspirates or blood were treated by
Hypaque Ficoll separation. Cell suspensions from solid or liquid materials were
fixed in 96% ethanol. For staining cells were washed and suspended in a solution of
combined ethidium bromide and mithramycin fluorescent dyes. Processing and
staining procedures have been described in detail for solid tumors [11, 43] and blood
or bone marrow [11, 16, 18]. As a ploidy standard nucleated blood cells of normal
donors were admixed in two different concentrations and stained together with the

---

1 Medizinische Klinik und Poliklinik der Universität, Albert-Schweitzer-Str. 33, D-4400 Münster, FRG

Tumor Aneuploidy
Büchner et al.
© Springer-Verlag: Berlin Heidelberg 1985

sample cells. Flow through measurement was carried out by ICP flow cytometers (Phywe AG., Partec AG.) [10]. DNA histograms were generally basing on 20 000 to some 100 000 measured cells. Coefficients of variation for $G_{1/0}$ cells reflecting measurement accuracy ranged from 1.3–4.5% and were mostly at 2.5–4%. The grade of DNA-aneuploidy of a cell clone was defined by the DNA-index [2, 3, 19] with 1.0 for normal diploid cells in $G_{1/0}$ phase.

Results

The data obtained by the cooperative study with respect to the above mentioned questions were as follows:

Incidence of DNA aneuploidy and tumor type: The incidences found in different malignancies are listed in Table 1. Three groups of tumors can be distinguished representing 1. carcinomas, melanomas, sarcomas and testicular tumors with 75–95%, also including myelomas with 65%, 2. acute leukemias with 40% aneuploidies and 3. basal cell skin carcinomas and congenital melanocytic nevi with 18.7 and 8.7% aneuploidies.

Grade of DNA aneuploidy and tumor type: Table 2 gives the DNA indices found in different malignancies. Thus, three groups are distinguishable:

1. carcinomas, melanomas and sarcomas with DNA indices mainly from 1.0 to 2.0 with a mean value of 1.75 and minor portions revealing DNA indices somewhat below or markedly above this range (Fig. 1),
2. testicular tumors (Fig. 2) spreading round a "triploid" value of 1.5,
3. hematological malignancies including myelomas and acute leukemias with DNA indices generally below 1.25

Typical DNA histograms for solid tumors and myelomas are shown in Figs. 3 and 4. The detection of low-grade DNA aneuploidy by admixed normal reference cells in acute leukemias is demonstrated in Fig. 5.

Table 1. Incidence of DNA-aneuploidy in the different malignances studied

Tumor	n	DNA-aneuploidy in	
		n	%
Malignant melanomas	721	548	76.0
Congen. melanocytic nevi	46	4	8.7
Squamous cell skin carcinomas	298	245	82.2
Basal cell carcinomas	509	95	18.7
Breast carcinomas	100	74	74.0
Colon carcinomas	53	43	81.1
Lung carcinomas	41	34	82.9
Testicular tumors	83	78	94.0
Soft tissue sarcomas	54	51	94.5
Osteosarcomas	16	13	81.3
Acute leukemias	394	161	40.9
Myelomas	37	24	64.9

42

Table 2. Grades of DNA-aneuploidy expressed by DNA-indices for the cases with DNA-aneuploidy found in the different malignancies

Tumor	n Aneuploid	DNA-Index	
		Mean	Range
Malignant melanomas	548	1.65	1.05–4.55
Congen. melanocytic nevi	4	1.50	1.30–1.70
Squamous cell skin carcinomas	245	1.82	1.05–3.20
Basal cell carcinomas	95	1.95	1.10–2.40
Breast carcinomas	74	1.67	1.05–2.90
Colon carcinomas	43	1.58	1.06–3.64
Lung carcinomas	34	1.55	0.75–3.40
Testicular tumors	78	1.54	1.05–3.0
Soft tissue carcinomas	51	1.68	1.10–2.60
Osteosarcomas	13	1.70	1.15–3.00
Acute leukemias	161	1.11	0.88–1.83
Myelomas	22	1.20	1.07–2.00

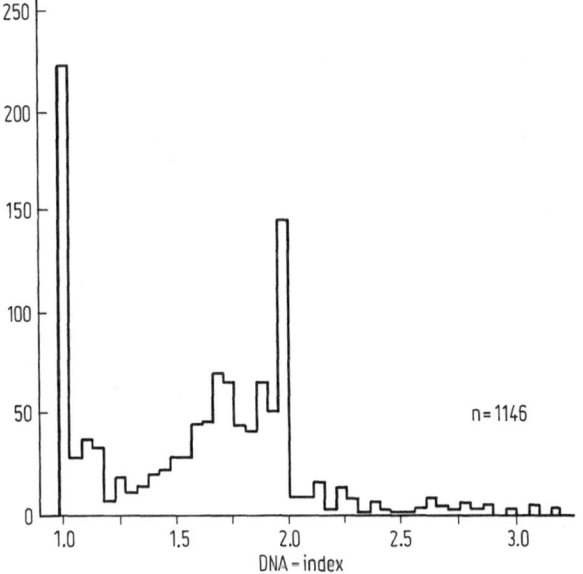

Fig. 1. Distribution of DNA-indices in various malignant non testicular solid tumors

Multiple clone DNA aneuploidy and tumor type: DNA histograms with multiple aneuploid clones are shown in Figs. 6–8 for acute leukemias and solid tumors. The frequency of multiple clone DNA aneuploidy was found lower in acute leukemias (6%) than in solid tumors (28%–44%, p < 0.05).

DNA aneuploidy and dignity: Table 3 compares various microscopically benign lesions and cell preparations with various material containing malignant cells by microscopy. Whereas DNA aneuploidy was found in the vast majority of malignant

43

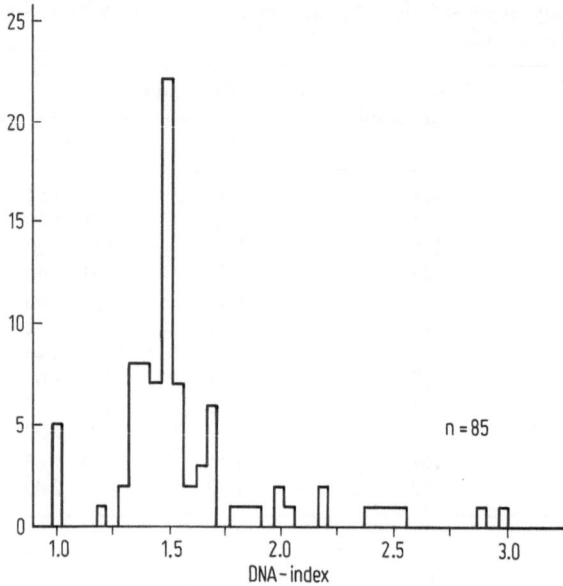

n = 85

Fig. 2. Distribution of DNA-indices in testicular tumors

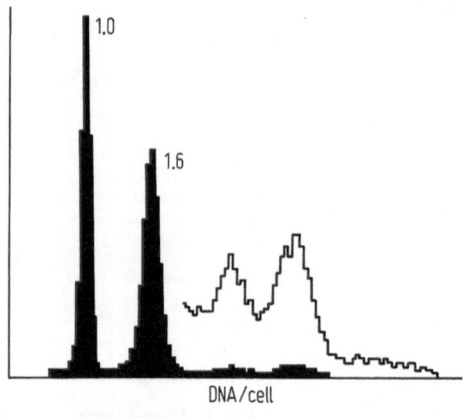

Fig. 3. DNA-aneuploidy in a lung carcinoma showing a DNA stemline at 1.6. The distribution of proliferating cells of both, the normal residual tissue and the malignant clone are overlappingly spanned out at higher DNA values

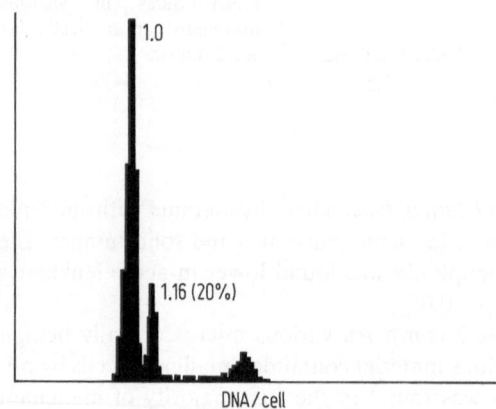

Fig. 4. A small portion of myeloma cells in the bone marrow detectable by its typically low-grade DNA-aneuploidy with a DNA-index of 1.16

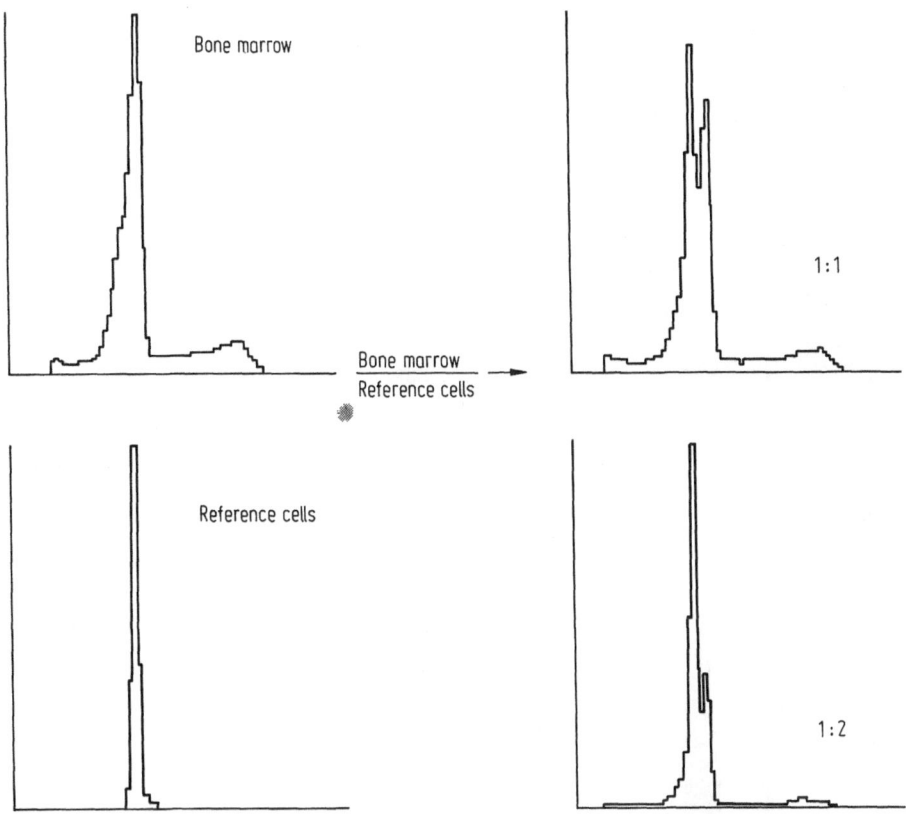

Fig. 5. DNA-histogram of a leukemic bone marrow showing a low-grade DNA-aneuploidy appearing as an asymmetry of the $G_{1/0}$ peak (left). By admixture of normal diploid reference cells in two different concentrations the aneuploid clone is identified with a DNA-index of 1.07

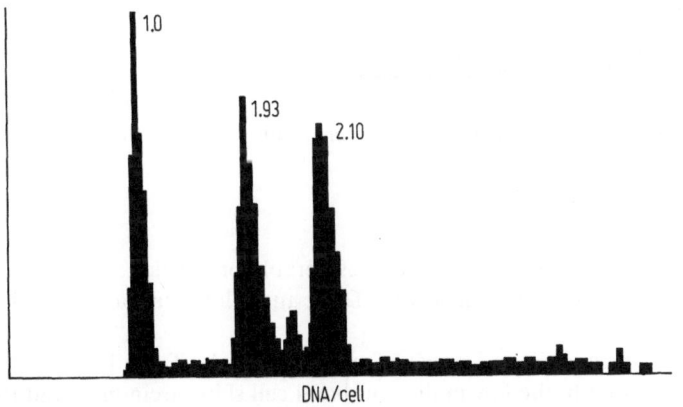

Fig. 6. Exceptional multiple clone DNA-aneuploidy in a case of AML with two near "tetraploid stemlines"

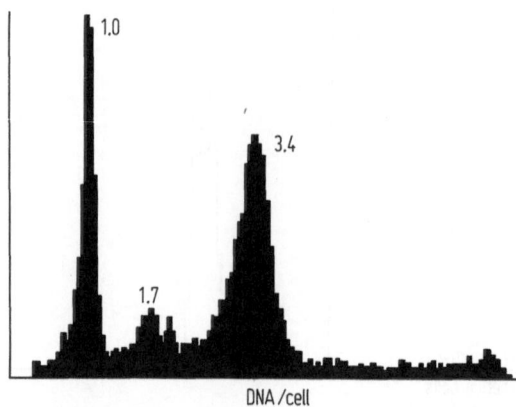

Fig. 7. Multiple clone DNA-aneuploidy in a lung carcinoma with double the DNA-index of the second as of the first stemline and proliferating cells of the second stemline (DNA-index 3.4) right from this peak

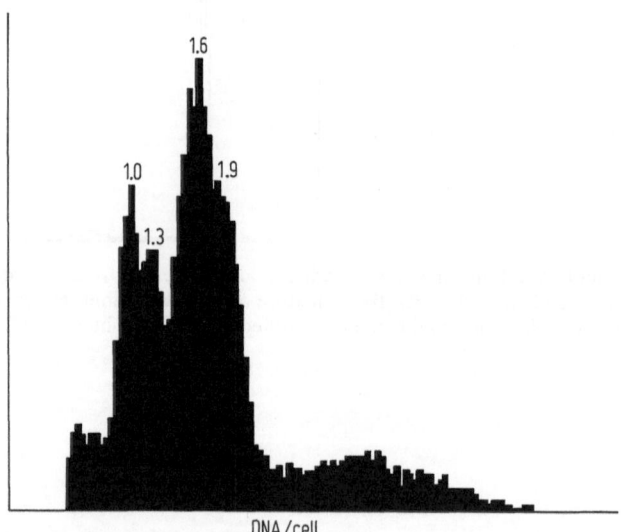

Fig. 8. Multiple clone DNA aneuploidy in an osteosarcoma with 3 distinguishable aneuploid clones at DNA indices 1.3, 1.6 and 1.9

frozen sections and pleural effusions and in a considerable proportion of acute leukemic bone marrows, there was no falsepositive DNA aneuploidy in the 776 benign samples.

Incidence of DNA aneuploidy and grade of malignancy: As Table 1 shows the DNA aneuploidy rate of both, the low malignant basal cell skin carcinomas and the potentially malignant congenital melanocytic nevi, is significantly lower when compared to either most solid tumors or acute leukemias.

46

Table 3. Comparison of DNA-aneuploidy of microscopically malignant samples to benign samples and lesions

Lesions and materials	n	DNA-aneuploidy	
		n	%
Acquired nevocellular nevi	376	0	0
Various benign skin lesions	207	0	0
Frozen sections			
Microscopically benign	45	0	0
Microscopically malignant	42	32	76.2
Pleural effusions			
Microscopically benign	34	0	0
Microscopically malignant	19	13	68.4
Bone marrow samples			
Microscopically benign	114	0	0
Acute leukemias	394	161	40.9

Table 4. Correlation of DNA-aneuploidy incidence to tumor thickness in primary melanomas (p=0.001)

Tumor thickness	n	DNA-aneuploidy in
<0.75 mm	64	31.3%
0.75–1.5 mm	57	52.6%
1.5–3.0 mm	51	60.8%
>3.0 mm	58	72.4%

Table 5. Correlation of DNA-aneuploidy incidence to surface area of primary melanomas (p=0.001)

Tumor surface area	n	DNA-aneuploidy in
<100 mm²	52	34.6%
100–200 mm²	53	45.3%
200–400 mm²	61	52.5%
>400 mm²	57	86.0%

Incidence of DNA aneuploidy and tumor stage: In 230 metastases forming melanomas, primary tumors showed a 45.3% DNA aneuploidy rate vs. 86.2% for the 491 metastases measured (p=0.001). There was a significant positive correlation for either the thickness (Table 4) or the surface area (Table 5) with the DNA aneuploidy rate in the primary lesions.

Incidence of DNA aneuploidy and prognosis: As Table 6 shows, DNA aneuploidy of totally resected primary melanomas, occurred with a significantly higher frequency of metastases after one and two year follow-ups.

In acute leukemias DNA-aneuploidy was compared with response and remission duration. There was no difference in remission rates between patients with and without DNA-aneuploidy neither in AML nor ALL. In the multicenter study BFM 79/81 on childhood ALL the probability of continuous complete remission after 5 years was higher for children with than for children without DNA aneuploidy (p=0.53). In the Münster study (78/81) on adult AML, patients with DNA aneuploidy have shown a higher probability of continuous complete remission after 4 years than patients without DNA aneuploidy (n.s.)

Discussion

DNA aneuploidy detected by FCM and its differential pattern according to the tumor type, tumor stage and prognosis was investigated in a multi-institutional study representing the largest set of malignancies and number of cases published so far.

Comparing 583 benign skin lesions and 193 microscopically benign frozen sections, pleural effusions and bone marrow samples with 2349 cases of malignant diseases and 64 microscopically malignant frozen sections and pleural effusions DNA aneuploidy was confirmed as a definite malignancy-specific cell marker, thus providing a reliable basis for diagnostic use in addition to morphology.

For the incidence of DNA aneuploidy we found 3 markedly different levels (Table 1), the highest one representing various solid tumors of which we could analyse 8 different entities with a DNA-aneuploidy rate of about 75% to 95%. The results are in agreement with findings in carcinomas and sarcomas by others (see 3) and confirm two previous studies on myelomas [8, 21] with a frequency of DNA aneuploidy of 65% on the same high level.

An intermediate level of about 40% aneuploidies is represented by acute leukemias. This result, basing on a representative number of 394 cases, high resolution fluorometry and standardised use of normal donor blood reference cells, considerably exceeds the rates found by other groups [3, 22], except one report on 40%. DNA aneuploidies in childhood ALL [37], however, not restricted to untreated ALL. Compared with cytogenetic findings, a 40% DNA aneuploidy rate is rather in agreement with the frequency of numerical chromosome aberrations in acute leukemias [27, 28, 38, 41].

A low level of DNA aneuploidy incidence (Table 1) is marked by the low grade malignant basal cell skin carcinomas (19%) and the potentially malignant congenital melanocytic nevi (9%) [36]. Low grade malignant lymphomas with 17% aneuploidies [4] are ranging on the same low level.

For the grade of DNA aneuploidy (Table 2) expressed by DNA index, we could distinguish between 3 groups of malignancies. The first group includes all non-testicular solid tumors and shows a broad distribution predominantly between DNA-indices of 1.0 and 2.0 with a minor portion somewhat below or markedly above this range (Fig. 1) as was also shown by others [3].

The second group is formed by testicular tumors with DNA indices distributed around a maximum of 1.5 (Fig. 2).

In the third group DNA indices cluster in the range below 1.2. This group includes hematological malignancies like acute leukemias, myelomas and also malignant lymphomas as shown by the Houston group [4].

Early animal experiments on chemical carcinogenesis have shown that DNA aneuploidy as detected by single cell cytophotometry does not appear during the stage of carcinogenesis but only in later stages of tumor progression as a secondary effect [12]. Endomitotic polyploidisation may be a major common mechanism during the evolution of aneuploid DNA stemlines. This is supported by the peak at the DNA index of 2.0 in the distribution of aneuploid solid tumors (Fig. 1) and is also suggested by the 2:1 ratio of DNA indices for multiple clone aneuploidy typical for various malignancies in present study. A similar mechanism may lead to the peak at the DNA index of 1.5 in tumors of the testis physiologically containing haploid and diploid cells. Progressive loss of chromosomes after polyploidisation may lead to new "hypotetraploid" or "hyperdiploid" stemlines. This could explain the incidence of clones with DNA-indices of 2.0 in early stages of bladder carcinomas and the shift to lower values in higher tumor stages and histologic grades [39].

There is only little understanding about the strong limitation of aneuploidy to low DNA indices in the hematological malignancies. It remains unsettled whether special properties of the blood cell system or dynamics of the diseases account for the differential pattern of the grade of DNA aneuploidy as compared with solid tumors.

In addition, the present study has given some evidence for correlations between DNA aneuploidy and malignancy grade, tumor stage and prognosis.

The first evidence was found in the observation of low DNA aneuploidy rates in the potentially malignant congenital melanocytic nevi [36] as was also shown for chronic ulcerative colitis by others [13] and in the low-grade malignant basal cell carcinomas (Table 1) corresponding to findings in low-grade malignant lymphomas [4].

The second evidence is basing upon our melanoma data [33] showing higher DNA aneuploidy rates in metastases than in primary tumors and a positive correlation of DNA aneuploidy rates with the thickness and surface area of primary tumors (Tables 4 and 5). In breast cancer DNA aneuploidy was found to correlate with undifferentiated morphology, lack of estrogene receptors [24] and higher proliferation intensity [26].

The coincidence of higher metastases formation rate with DNA aneuploidy in primary melanomas (Table 6) gives a strong evidence for a prognostic significance of this cell marker.

In contrast, the low-grade aneuploidy of acute leukemias rather seems to predict for longer remission duration [18, 25] as also shown by others, using single cell cytophotometry [20], FCM [22] or chromosome analysis [35, 38]. As a speculation the low grade DNA aneuploidies as present in the blood cell system may predominantly

Table 6. Frequency of metastases formation in resected primary melanomas with and without DNA-aneuploidy (p = 0.05)

	n	1 year-	2 years follow-up
Primary tumors			
with DNA-aneuploidies	125	17 (13.6%)	21 (16.8%)
without DNA-aneuploidies	105	4 (3.8%)	6 (5.7%)

occur with functional inferiority, whereas the evolution of functionally superior aggressive clones may be possible only in solid tumors. As shown in childhood neuroblastoma, a solid tumor type high grade DNA aneuploidy may, however, be associated with better response to chemotherapy [23]. Resuming the controversial data on DNA aneuploidy and prognosis a differential prognostic significance according to natural history, response to therapy and different tumor types has to be discussed.

References

1. Atkin NB, Mattinson G, Baker MC (1966) A comparison of the DNA content and chromosome number of fifty human tumors. Br J Cancer 20:87–101
2. Barlogie B, Göhde W, Johnston DA, Smallwood L, Schumann J, Drewinko B, Freireich EJ (1978) Determination of ploidy and proliferative characteristics of human solid tumors by pulse cytophotometry. Cancer Res 38:3333–3339
3. Barlogie B, Raber MN, Schumann J, Johnson TS, Drewinko B, Swartzendruber DE, Göhde W, Andreeff M, Freireich EJ (1983) Flow cytometry in clinical cancer research. Cancer Res 43:3982–3997
4. Barlogie B (1985) Flow cytometry as a diagnostic and prognostic tool in cancer medicine. In: Büchner Th, Bloomfield CD, Hiddemann W, Hossfeld D, Schumann J (eds) Tumor Aneuploidy. Springer, Berlin Heidelberg New York Tokyo
5. Bloomfield CD, Lindquist LL, Arthur D, McKenna RW, LeBien TW, Peterson BA, Nesbit ME (1981) Chromosomal abnormalities in acute lymphoblastic leukemia. Cancer Res 41:4838–4843
6. Büchner Th, Dittrich W, Göhde W (1971) Die Impulscytophotometrie in der hämatologischen Cytologie. Klin Wschr 49:1090–1092
7. Büchner Th, Dittrich W, Göhde W (1972) Automatische DNS-Messungen zur Zellkinetik von Leukämien mit Hilfe der Impulscytophotometrie In: Gross R, van de Loo J (eds) Leukämie. Springer Berlin Heidelberg New York, p 205–208
8. Bunn PA, Krasnow S, Makuch RW, Schlam ML, Schechter FP (1982) Flow cytometric analysis of DNA content of bone marrow cells in patients with plasma cell myeloma: clinical implications. Blood 59:528–533
9. Van Dilla MA, Trujillo TT, Mullaney PF, Coulter JR (1969) Cell microfluorometry: a method for rapid fluorescence measurement. Science 163:1213–1214
10. Göhde W, Dittrich W (1971) Impulsfluorometrie – ein neuartiges Durchflußverfahren zur ultraschnellen Mengenbestimmung von Zellinhaltsstoffen. Acta Histochem 10:429–437
11. Göhde W, Schumann J, Büchner Th, Otto F, Barlogie B (1979) Pulse cytophotometry: Application in tumor cell biology and clinical oncology. In: Melamed MR, Mullaney PF, Mendelsohn ML (eds) Flow cytometry and sorting. John Wiley and Sons Inc., New York, p 599–620
12. Grundmann E (1954) DNS-Messungen an den Zellkernen der Rattenleber nach partieller Hepatektomie und während der Carcinogenese. Klin Wochenschr 32:1023
13. Hammarberg C, Slezak P, Tribukait B (1984) Early detection of malignancy in ulcerative colitis. A flow-cytometric DNA-study. Cancer 53:291–295
14. Hossfeld DK (1985) Zytogenetik maligner Erkrankungen In: Gross R, Schmidt CG (eds) Klinische Onkologie. Georg Thieme Verlag, Stuttgart
15. Hiddemann W, Meister R, von Bassewitz DB, Wörmann B, Büchner Th (1984) DNS-Aneuploidie: Ein hochspezifischer Marker für Malignität in der Diagnostik von Pleuraergüssen. Atemw Lungenkrankh 7:328–330
16. Hiddemann W, Büchner Th, Wörmann B, von Bassewitz DB, Müller KM, Ritter J, Hauss J, Kleinemeier HJ, Grundmann E (1985) Incidence and heterogeneity of DNA aneuploidies in solid tumors and acute leukemias. In: Büchner Th, Bloomfield CD, Hiddemann W, Hossfeld D, Schumann J (eds) Tumor aneuploidy. Springer, Berlin Heidelberg New York Tokyo
17. Hiddemann W, Springefeld R, Büchner Th (1984) DNS-Aneuploidien bei multiplem Myelom. Inzidenz und klinische Bedeutung. Verh Dtsch Ges Inn Med

18. Hiddemann W, Wörmann B, Büchner Th (1985) DNA aneuploidies in adult patients with acute myeloid leukemia – incidence and relation to patients characteristics and prognosis. Submitted for publication
19. Hiddemann W, Schumann J, Andreeff M, Barlogie B, Herman CJ, Leif RC, Mayall BH, Murphy RF, Sandberg AA (1984) Convention on nomenclature for DNA cytometry. Cytometry 5:445–446
20. Lampert F (1969) Quantitative Zytologie der akuten Leukämie im Kindesalter. Fortschr Medizin 87:83–126
21. Latreille J, Barlogie B, Dosik G, Johnston DA, Drewinko B, Alexanian R (1980) Cellular DNA content as a marker of human multiple myeloma. Blood 55:403–408
22. Look AT, Melvin SL, Williams DL, Brodeur GM, Dahl GV, Kalwinsky DJ, Murphy SB, Mauer AML (1982) Aneuploidy and percentage of S-phase cells determined by flow cytometry correlate with cell phenotype in childhood acute leukemia. Blood 60:959–967
23. Look AT, Hayes FA, Nitschke R, McWilliams NB, Green AA (1984) Cellular DNA content as a predictor of response to chemotherapy in infants with unresectable neuroblastoma. N Engl J Med 311:231–235
24. Olsezewski W, Darzynkiewicz Z, Rosen PP, Schwartz MK, Melamed MR (1981) Flow cytometry of breast carcinoma: I. relation of DNA ploidy level to histology and estrogen receptor. Cancer 48:980–984
25. Ritter J, Hiddemann W, Wörmann B, Schellong G, Büchner Th (1985) DNA-aneuploidy in childhood acute lymphoblastic leukemia as detected by flow cytometry: relation to phenotype, presentation, features and prognosis. In: Büchner Th, Bloomfield CD, Hiddemann W, Hossfeld D, Schumann J (eds) Tumor Aneuploidy. Springer, Berlin Heidelberg New York Tokyo
26. Raber MN, Barlogie B, Latreille J, Bedrossian C, Fritsche H, Blumenschein G (1982) Ploidy, proliferative activity and estrogen receptor content in human breast cancer. Cytometry 3:36–41
27. Rowley JD (1978) The cytogenetics of acute leukemia. Clin Haematol 7:385–406
28. Rowley JD (1981) Chromosome studies in children and adults with leukemia. In: Neth R, Gallo RC, Graf TG, Mannweiler K, Winkler K (eds) Modern Trends in Human Leukemia IV. Springer, Berlin Heidelberg New York, p 64–70
29. Rowley JD, Golomb HM, Vardiman JW (1981) Nonrandom chromosome abnormalities in secondary acute leukemia: Relevance to etiology of ANLL de novo. Blood 58:759–767
30. Sandberg AA (1980) The chromosome in human cancer and leukemia. Elsevier North-Holland, New York
31. Sandritter W (1981) Quantitative pathology in theory and practice. – The Maude Abbott Lecture. Pathol Res Pract 171:2–21
32. Schumann J, Zante J, Göhde W (1978) Aneuploidies in solid human tumors. In: Lutz D (ed) Third International Symposium on Pulse Cytophotometry. European Press Medikon, Ghent
33. Schumann J, Tilkorn H, Göhde W, Ehring F, Straub C (1981) Zytogenetik maligner Melanome. Verh Dtsch Dermatol Ges, Hautarzt [Suppl] V, 32:62–66
34. Secker-Walker LM, Lawler SD, Hardisty RM (1978) Prognostic implications of chromosomal findings in acute lymphoblastic leukemia at diagnosis. Br Med J 2:1529–1530
35. Secker-Walker LM, Swansbury GJ, Hardisty RM, Sallan SE, Garson OM, Sakurai M, Lawler SD (1982) Cytogenetics of acute lymphoblastic leukaemia in children as a factor in the prediction of longterm survival. Br J Haematol 52:389–399
36. Stenzinger W, Suter L, Schumann J (1984) DNA aneuploidy in congenital melanocytic nevi: Suggestive evidence for premalignant changes. J Invest Dermatol 82:569–572
37. Suarez C, Miller D, Steinherz P, Andreeff M (1981) Flow cytometry for DNA and RNA determination in 107 cases of childhood acute lymphoblastic leukemia (ALL): correlation with FAB classification. Proc Am Ass Cancer Res 22:770
38. The Third International Workshop on Chromosomes in Leukemia (1981) Cancer Genet Cytogenet 4:96–137
39. Tribukait B (1984) Flow cytometry in surgical pathology and cytology of tumors of the genito-urinary tract. In: Koss LG, Coleman M (eds) Advances in Clinical Cytology, Vol. 2, Masson Publ USA Inc, p 163–189

40. Vindelow LL, Hansen HH, Christensen J, Spang-Thomsen M, Hirsch FR, Hansen M, Nissen NI (1980) Clonal heterogeneity of small-cell anaplastic carcinoma of the lung demonstrated by flow-cytometric DNA analysis. Cancer Res 40:4295–4300
41. Williams DL, Tsiatis A, Brodeur GM, Look AT, Melvin SL, Bowman WP, Kalwinsky DK, Rivera G, Dahl GV (1982) Prognostic importance of chromosome number in 136 untreated children with acute lymphoblastic leukemia. Blood 60:864–871
42. Wörmann B, Hiddemann W, Ritter J, Henze G, Langermann HJ, Büchner Th (1985) Incidence and distribution of DNA-aneuploidies in acute leukemia. In: Büchner Th, Bloomfield CD, Hiddemann W, Hossfeld D, Schumann J (eds) Tumor Aneuploidy. Springer, Berlin Heidelberg New York Tokyo
43. Zante J, Schumann J, Barlogie B, Göhde W, Büchner Th (1976) New preparating and staining procedures for specific and rapid analysis of DNA-distributions. In: Göhde W, Schumann J, Büchner Th (eds) 2nd Int. Symp, Pulse Cytophotometry, European Press, Medikon, Ghent, p 97–106

Incidence and Prognostic Significance of DNA Aneuploidy in Childhood Acute Lymphoblastic Leukemia

B. Wörmann[1], W. Hiddemann, J. Ritter, G. Henze, H.-J. Langermann, U. Kaufmann, G. Schellong, H. Riehm, Th. Büchner

Introduction

Recent improvements in cytogenetic technology provided the means to identify chromosomal abnormalities in the majority of children with acute lymphoblastic leukemia (ALL) [1–8]. In addition to numeric aberrations, detected in 23% to 39% of all cases with evaluable metaphases [1–4], specific nonrandom translocations have been described [5–8]. Cytogenetic studies have also emphasized the clinical significance of chromosomal aberrations in relation to previously identified risk factors such as male sex, high white blood cell count (WBC), age and T-cell phenotype [2, 3]. The applicability of cytogenetic techniques, however, is limited by the laborious and time consuming processing of samples and requires cell proliferation in vitro for the evaluation of metaphases.

In contrast, flow cytometry (FCM) provides a rapid and reliable analysis of the cellular DNA content, although the yield of detectable aneuploidies has been in the range of only 15% to 29% of the cases [9, 10]. Recent developments in flow systems and preparation techniques improved the accuracy of FCM measurements significantly. The present study was carried out on 100 children with ALL in order to determine:

1. the incidence of DNA aneuploidies in childhood ALL detectable by FCM,
2. the relation of DNA aneuploidies to patient characteristics such as sex, age, white blood cell count (WBC), immunologic markers and a risk factor recently defined [11].
3. the prognostic significance of DNA aneuploidies for therapeutic response and remission duration.

Material and Methods

Patients and Material

One hundred children with previously untreated ALL were entered into the study from three pediatric centers in West Germany between July 1979 and January 1982. Of them, 64 children, recruited between July 1979 and March 1981, were treated according to the BFM protocol 1979/81 which has been evaluated recently [12]. The other 36 patients in whom treatment was started after March 1981 were entered into

---

1 Dept. of Laboratory Medicine and Pathology, Box 609, Mayo Memorial Building, 420 Delaware Street S.E., Minneapolis, Minnesota 55455, USA

Tumor Aneuploidy
Büchner et al.
© Springer-Verlag: Berlin Heidelberg 1985

Table 1. Clinical charateristics of 100 children with ALL

ALL subtype	n	Median age (years)	Boys	Girls	WBC (mean)	Risk factor (mean)
T-ALL	14	7.8	10	4	180,504	1.39
B-ALL	3	11.0	3	0	17,900	1.26
Non-T/non-B ALL	83	5.4	46	37	44,378	0.95
All	100	5.9	59	41	47,976	1.02

a new regimen and have not yet been evaluated for clinical response, but were included for the analysis of FCM measurements and pretherapeutic data. Patients characteristics are listed in Table 1. Patients were classified according to a risk index based on the analysis of preceding studies [13]. Children with a risk index $\geq$ 3 received an intensified reinduction treatment subsequently referred to as regimen B. The standard risk protocol is referred to as A.

More recently a modification of the risk index was developed taking into account high WBC, splenomegaly and hepatomegaly. This so-called risk factor is of even higher prognostic value und is used for the treatment stratification in the present BFM study [11]. Risk factors were calculated retrospectively for the patients of the BFM study 1979/81 and used for the comparison with FCM aneuploidies and additional patient characteristics.

The diagnosis of ALL was based on the morphologic evaluation of Wright-Giemsa stained bone marrow smears and additional cytochemical procedures for the exclusion of nonlymphoblastic leukemias. T-ALL was identified by sheep erythrocyte-rosette formation at 4 °C and at 37 °C, B-ALL by the presence of surface immunoglobulins.

Flow Cytometry

For FCM analysis 0.5–1.0 ml of bone marrow and/or 1–5 ml of peripheral blood were aspirated into a syringe containing preservative free heparin (1000 U/ml) or EDTA (10 µg/ml) for anticoagulation. Samples from Münster were prepared within one hour after aspiration. Samples from Berlin and Giessen were sent by mail arriving one to three days after aspiration. 15% of the specimes obtained by mail could not be evaluated because of clotting or autolysis.

Bone marrow aspiration and blood specimens were subjected to Ficoll Hypaque gradient separation (density 1.078 g/ml) at 1000 g for 20 minutes at room temperature. The cell layer was removed, cells were washed twice in HBSS and fixed in 96% ethanol and stored at 4 °C. Mononuclear cells from healthy blood donors enriched by leukapheresis served for reference measurements and were prepared in the same manner. For FCM analysis the fixed cells were centrifuged again at 1000 g for 5 minutes, the pellet was resuspended in 1 ml of a 0.5% pepsin-HCl solution (pepsin – Merck 1000 U/g) for one minute and stained with Ethidium Bromide and Mithramycin in combination (Ethidium Bromide – Serva, 10 mg/1000 ml Tris buffer pH 7.5; Mithramycin – Serva, 25 mg/1000 ml Tris buffer pH 7.5) [14, 15]. For each histogram 20,000 to 50,000 cells were measured using an ICP 11 (Phywe AG).

The determination of the DNA ploidy was based on reference measurements with mononuclear cells from healthy blood donors mixing patient and reference cells at ratios of $1:1$ and $1:2$. The ratio of the $G_{0/1}$ peaks of the patient sample and the reference cells multiplied by 2 provides the DNA index. The DNA index of diploid cells is 1.0. Based on previous studies defining a margin of error of $\pm$ 3.0% for the determination of the DNA index [16], DNA indices of ≤ 0.96 and ≥ 1.04 are considered as aneuploid.

Statistics

Statistical evaluation included the life table analysis method for remission duration [17], the chi-square test and the parameter free comparison of independent data by Nemenyi [18]. All results were updated as of February 16, 1982.

Results

Incidence of Aneuploidies

Applying reference measurements to all 100 cases evaluated dominant aneuploid cell lines were detected in 37 patients (37%) (Fig. 1). The DNA index ranged from 0.88 to 2.0 with a mean of 1.18 and a median of 1.16 (Tables 2 and 3). Two patients showed hypodiploid cell lines, in one case a tetraploid aneuploidy was found. Two different cell lines were identified in the bone marrow of a two years old boy with a non-T/non-B ALL, 96% of the cells having a DNA index of 1.0, 4% being hyperdiploid with a DNA index of 1.77. Morphological and immunological analyses revealed only one cell type.

In 18 children with aneuploid leukemic blasts bone marrow aspirates were re-evaluated 4 weeks after initiation of therapy. Cytological examination did not reveal persistent leukemic blasts in any case. By FCM analysis no aneuploidies were found in either case. In relapse, two boys with initially diploid leukemic blasts revealed a unimodal distribution of diploid cells again.

One 12 years old girl with non-T/non B ALL had shown a unimodal DNA distribution in her blood prior to therapy which was identified by reference measure-

Table 2. Incidence of DNA aneuploidies in 100 children with ALL

	n	DNA aneuploidy	DNA index
Münster	71	27 (38%)	0.88–2.0
Berlin + Gießen	29	10 (34%)	0.96–1.38
All	100	37 (37%)	0.88–2.0

Table 3. Distribution of DNA indices in 100 children with ALL

DNA Index	≤ 0.96	1.00	$\geq 1.04 >$	$\leq 1.10 >$	$\leq 1.20 >$	$\leq 1.30 >$	$\leq 1.40 > > 1.40$
n	2	63		7	21	4	2	2

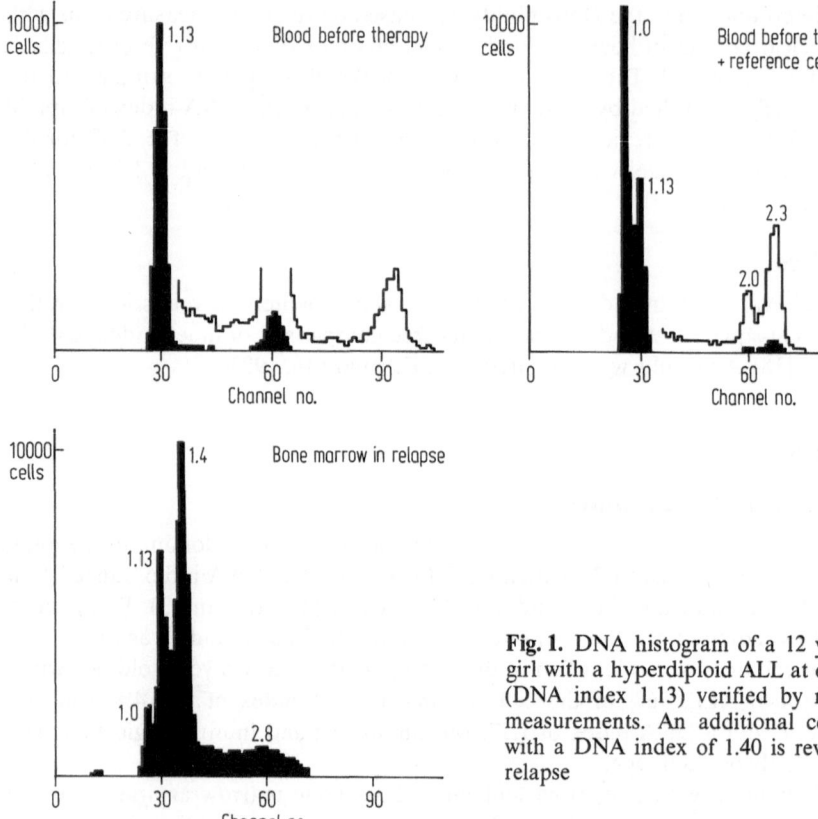

Fig. 1. DNA histogram of a 12 years old girl with a hyperdiploid ALL at diagnosis (DNA index 1.13) verified by reference measurements. An additional cell clone with a DNA index of 1.40 is revealed at relapse

ments as being hyperdiploid with a DNA index of 1.13 (Fig. 1). In relapse a new cell line with a DNA index of 1.40 was detected in addition to the reappearance of the previously diagnosed aneuploid clone.

Aneuploidy in Relation to Patient Characteristics

1. Immunologic Markers

Of 14 patients with T-ALL only one (7%) had an aneuploidy with a DNA index of 1.06. This boy is the only patient with T-ALL in the analysed group who relapsed up to now. One out of three patients with B-ALL was aneuploid with a DNA index of 1.08. Of 83 patients with non-T/non-B ALL 35 (43%) revealed aneuploid cell lines with DNA indices between 0.88 and 2.0. The difference in the incidence of aneuploidies in T-ALL and non-T/non-B ALL is highly significant (p < 0.001) (Table 4).

2. Sex and Age

59 patients were male, 41 female. 17 boys (29%) had expressed aneuploidies, 20 girls (49%) had aneuploid cell lines (p < 0.05). Without considering the patients with

Table 4. Incidence of DNA aneuploidies in 100 children with ALL according to immunologic subgroups

	n	DNA aneuploidy	DNA index	p
T-ALL	14	1 (7%)	1.06	
Non-T/ non-B ALL	83	35 (43%)	0.88–2.0	<0.001
B-ALL	3	1	1.08	

T-ALL the majority of whom is male and diploid, the difference remains significant at p<0.05.

The age of patients with non-T/non-B ALL (mean 5.4 years) is significantly lower than for patients with T-ALL (mean 7.8 years) and B-ALL (mean 11.0 years) (p<0.01). Among the children with non-T/non-B ALL no differences in age were found for patients with (mean 5.5 years) or without (mean 5.4 years) aneuploidies.

3. White Blood Cell Count (WBC)

The distribution of WBCs in schildren with T-ALL and with diploid and aneuploid non-T ALL is shown in Fig. 2. Patients with aneuploid non-T ALL have a significantly lower WBC with a mean of 12,536 as compared to patients with diploid non-T ALL (mean 65,636) and children with T-ALL (mean 180,504) (p<0.05) (Table 5). The only patient with an aneuploid T-ALL had a WBC of 169,000 which is not different from the mean value in T-ALL.

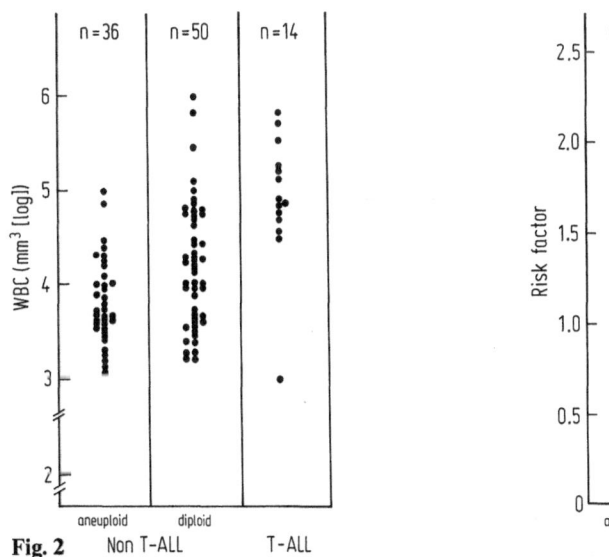

Fig. 2

Fig. 3

Fig. 2. Distribution of WBC in 100 children with T-ALL and diploid and aneuploid non-T ALL

Fig. 3. Distribution of risk indices in 100 children with T-ALL and diploid and aneuploid non-T ALL

4. Risk Factor

Children with T-ALL have a significantly higher risk factor (mean 1.39) than patients with non-T ALL ($p < 0.05$) (Fig. 3, Table 6). The difference between diploid and aneuploid non-T ALL is not significant.

Table 5. WBC in 100 children with ALL according to immunologic markers and DNA ploidy

ALL subtype	n	Mean WBC	S_{dev}	p
T-ALL	14	180,504	±215,350	<0.05
Diploid non-T ALL	50	65,636	±181,903	
Aneuploid non-T ALL	36	12,536	± 19,065	<0.05

Table 6. Risk factors in 100 children with ALL according to immunological markers and DNA ploidy

ALL subtype	n	Mean risk factor	S_{dev}	
T-ALL	14	1.39	±0.52	
Diploid non-T/ALL	50	1.02	±0.48	$p < 0.$
Aneuploid non-T/ALL	36	0.85	±0.48	

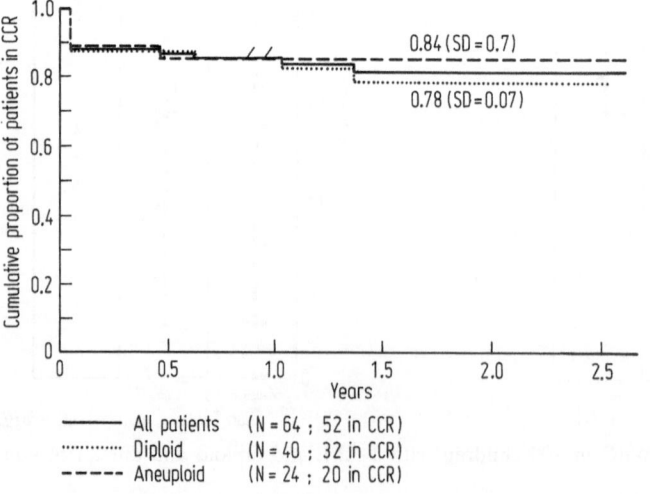

Fig. 4. Life table analysis for 64 patients of the BFM study 1979/81 comparing diploid and aneuploid ALL

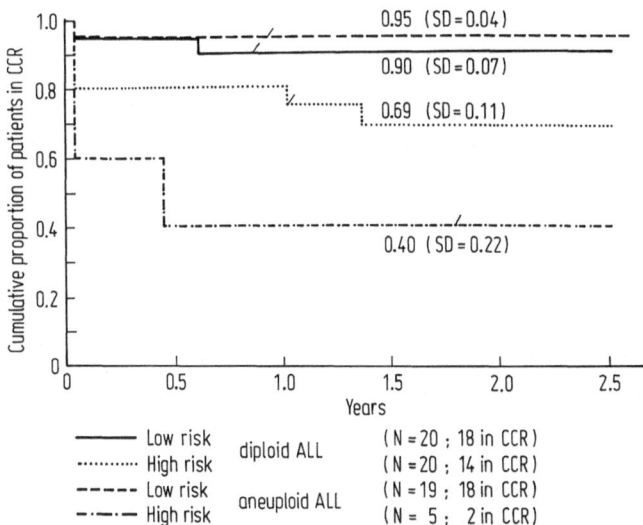

Fig. 5. Life table analysis for 64 children of the BFM study 1979/81 comparing diploid and aneuploid ALL with high and standard risk indices

Prognostic Significance of Aneuploidies

Life table analysis for the 64 patients on the BFM study 1979/81 shows a cumulative proportion of patients in continuous complete remission at 32 months of 81%. The percentage of children with aneuploid ALL in CCR (84%) is slightly but not significantly higher as compared to diploid ALL (78%) (Fig. 4). Life table analysis according to the standard risk protocol A and the high risk protocol B shows that the proportion of children in CCR is lowest for high risk patients with diploid and aneuploid stemlines (69% and 40%) and highest in patients with low risk and aneuploid leukemic blasts (95%). Protocol B was applied to 5 of 19 patients with aneuploid ALL (18.5%) while 20 out of 40 with diploid ALL (50%) required the intensified reinduction therapy (Fig. 5).

Discussion

Recent improvements of the technology of flow systems and of cell preparation techniques have resulted in a high accuracy and reliability of flow cytometric analyses of the cellular DNA content. In addition, systematic reference measurements with mononuclear cells from normal blood donors have provided the means to detect aneuploid cell lines in a large proportion of patients with acute leukemias [19, 20]. Based on previous studies a range of error of ± 3% for DNA index determinations was defined for the present study [16]. DNA indices ≤ 0.96 and ≥ 1.04 are therefore considered as representing aneuploid cell lines.

Applying these criteria, an incidence of 37% aneuploidies was found in 100 children with newly diagnosed acute lymphoblastic leukemia. These results considerably exceed previously reported data on FCM measurements in ALL [9, 10]. In large cytogenetic studies the percentage of numeric aberrations in ALL was between 23% and 39%. These data closely correspond to our findings [1–4].

Biclonal leukemias were observed in two patients, in one at the time of diagnosis, in the other at relapse when the reoccurence of a previously identified hyperdiploid cell clone (DNA index 1.18) was accompanied by the appearance of a second hyperdiploid leukemic cell line with a higher DNA index of 1.40. Similar data have been reported by Andreeff et al. [21], while Look et al. [10] found an even higher incidence of biclonal ALL in children combining flow cytometry and karyotype analysis.

Previous cytogenetic investigations have revealed a correlation between chromosomal aberrations and clinically defined risk factors in childhood ALL such as T-cell phenotype, high WBC and age. A significantly higher incidence of karyotype abnormalities in non-T/non-B ALL as compared to T-ALL was reported at the Third International Workshop on Chromosomes in Leukemia [3], a finding that could be confirmed in our study. Andreeff et al. [22] and Brodeur et al. [23] identified near-haploid ALL as a subtype with especially poor prognosis. In the present study no statistically significant difference in remission duration was found between patients with aneuploid and diploid ALL with a slight tendency towards longer remissions in patients with aneuploidy. This finding, however, can also be explained by the higher WBC in patients with diploid ALL, therefore carrying a higher risk factor than children with aneuploid ALL. WBCs were found to be highest in the subgroup of T-ALL with only one of 14 patients having an aneuploid cell line. It should be noted in this context, that the diagnosis of T-ALL per se does not establish a poor prognosis [13]. In general the prognostic significance of all so-called risk factors in childhood ALL seems to be largely dependent on the intensity of the applied therapeutic regimen.

For the BFM protocol 1979/81 patients younger than 2 years and older than 10 years had a worse prognosis than children between 2 and 10 years of age [12]. No correlation was found, however, between aneuploidy and age for patients within the non-T/non-B ALL group. Children with T-ALL were older, a previously described phenomenon unrelated to aneuploidy.

Aneuploidy is a highly specific marker for malignancy. Small numbers of leukemic cells can be recognized objectively when blasts are not yet identifiable by morphology as for example in therapy-induced bone marrow hypoplasia or during complete remission. The diagnostic potential of flow cytometry for the detection of aneuploid cells is limited by the resolution power of the flow system used and can certainly be further increased by additional technical improvements. It has to be considered, however, that even with high standard FCM analysis only changes in the total cellular DNA content can be identified by flow cytometry. That means that numeric aberrations are detectable while structural rearrangements cannot be identified. Cytogenetics, on the other hand, requires cell proliferation in vitro to gain evaluable metaphases. Hence the correlation of cytogenetic data with factors concerning clinical response and prognosis in ALL is based on a selected group of patients. Cases without cell growth in vitro are excluded.

The results of the present study clearly indicate that FCM analysis provides a rapid and reliable evaluation of cellular DNA content detecting aneuploidies in a large proportion of children with acute lymphoblastic leukemia.

References

1. Rowley JD (1978) The cytogenetics of acute leukemia. Clinics Haemat 7,2:385
2. Arthur DC, Bloomfield CD, Lindquist LL, Peterson BA, Nesbit ME (1981) Chromosome abnormalities in acute lymphoblastic leukemia (ALL): Frequency and clinical implications. Proc ASCO 22:345 (Abstr)
3. Bloomfield CD (1981) Clinical significance of chromosomal abnormalities in acute lymphoblastic leukemia (ALL). A preliminary report of the Third International Workshop on Chromosomes in Leukemia (TIWCL). Proc ASCO 22:228 (Abstr)
4. Kaneko Y, Rowley JD, Variakojis D, Chilcote RR, Check I, Sakurai M (1982) Correlation of karyotype with clinical features in acute lymphoblastic leukemia. Cancer Res 42,7:2918
5. Priest JR, Robison LL, McKenna RW, Lindquist LL, Warkentin PI, LeBien TW, Woods WG, Kersey JH, Coccia PF, Nesbit ME (1980) Philadelphia chromosome positive childhood acute lymphoblastic leukemia. Blood 56:12
6. Cimino MC, Rowley JD, Kinnealy A, Variakojis D, Golomb HM (1979) Banding studies of chromosomal abnormalities in patients with acute lymphoblastic leukemia. Cancer Res 39:227
7. Oshimura M, Freeman AI, Sandberg AA (1977) Chromosomes and causation of human cancer and leukemia. XXVI: Banding studies in acute lymphoblastic leukemia (ALL). Cancer 40:1161
8. Whang-Peng J, Knutsen T, Ziegler J, Leventhal B (1976) Cytogenetic studies in acute lymphoblastic leukemia: Special emphasis on long-term survival. Med Pediatr Oncol 2:333
9. Barlogie B, Latreille J, Freireich E, Fu CT, Mellard D, Meistrich M, Andreeff M (1980) Characterization of hematological malignancies by flow cytometry. Blood Cells 6:714
10. Look AT, Williams DL, Brodeur GM, Murphy SB, Dahl GV, Mauer AM (1982) Multiple stemlines in childhood acute lymphoblastic leukemia determined by flow cytometry (FCM). Proc AACR 23:33 (Abstr)
11. Langermann HJ, Henze G, Wulf M, Riehm HJ (1982) Abschätzung der Tumorzellmasse bei der akuten lymphoblastischen Leukämie im Kindesalter: Prognostische Bedeutung und praktische Anwendung. Klin Pädiatr 194:209
12. Henze G, Langermann HJ, Fengler R, Brandeis M, Evers KG, Gadner H, Hinderfeld L, Jobke A, Kornhuber B, Lampert F, Lasson U, Ludwig R, Müller-Weihrich S, Neidhardt M, Nessler G, Niethammer D, Rister M, Ritter J, Schaaff A, Schellong G, Stollmann B, Treuner J, Wahlen W, Weinel P, Wehinger H, Riehm J (1982) Bericht über die Therapiestudie BFM 1979/81 zur Behandlung der akuten lymphoblastischen Leukämie bei Kindern und Jugendlichen: Intensivierte Reinduktionstherapie für Patientengruppen mit unterschiedlichem Rezidivrisiko. Klin Pädiatr 194:195
13. Henze G, Langermann HJ, Kaufmann U, Ludwig R, Schellong G, Stollmann B, Riehm H (1981) Thymic involvement and initial white blood count in childhood acute lymphoblastic leukemia. Am J Ped Hemat Oncol 3:369
14. Zante J, Schumann J, Barlogie B, Göhde W, Büchner Th (1976) New preparation and staining procedures for specific and rapid analysis of DNA distributions. Pulse Cytophotometry, 2nd Int. Symp. In: Göhde W, Schumann J, Büchner Th (eds) European Press Ghent 97
15. Barlogie B, Spitzer G, Hart JS, Johnston DA, Büchner T, Schumann J, Drewinko B (1976) DNA histogram analysis of human hematopoietic cells. Blood 48:245
16. Wörmann B (1982) Intensivierte Induktionstherapie der akuten myeloischen Leukämie des Erwachsenen: Prognostische Bedeutung einiger zellulärer und zellkinetischer Parameter. Thesis, Münster
17. Cutler S, Ederer F (1958) Maximum utilisation of the life table method in analyzing survival. J. Chron. Dis. 4:699
18. Nemenyi P (1963) Distribution-free multiple comparisons. State University of New York, Downstate Medical Center
19. Barlogie B, Hittelman W, Spitzer G, Hart JS, Trujillo JM, Smallwood L, Drewinko B (1977) Correlation of DNA distribution abnormalities with cytogenetic findings in human adult acute leukemia and lymphoma. Cancer Res 37:4400

20. Hiddemann W, Wörmann B, Ritter J, Henze G, Langermann HJ, Kaufmann U, Schellong G, Riehm H, Büchner Th (1982) Diagnostik von Aneuploidien bei akuten Leukämien mittels Impulscytophotometrie (ICP): Häufigkeit und klinische Relevanz. Verh Dtsch Ges Inn Med 88:934
21. Andreeff M, Gee T, Mertelsmann R, McKenzie S, Steinmetz J, Chaganti R, Koziner B, Clarkson B (1980) Biclonal lymphoblastic and myeloblastic acute leukemia. Proc AACR 21:54
22. Andreeff M, Miller D, Steinherz P, Kempin S, Straus D, Clarkson B (1981) Hypodiploid acute lymphoblastic leukemia (ALL): A rare entity detected and monitored by flow cytometry. Proc AACR 22:44 (Abstr)
23. Brodeur GM, Williams DL, Look AT, Bowmann WR, Kalwinsky DK (1981) Near-haploid acute lymphoblastic leukemia: A unique subgroup with a poor prognosis? Blood 58:14
24. Sen L, Borella L (1975) Clinical importance of lymphoblasts with T markers in childhood acute leukemia. N Engl J Med 292:828

DNA Aneuploidy in Childhood Acute Lymphoblastic Leukemia as Detected by Flow Cytometry: Relation to Phenotype, Presentation Features and Prognosis

J. Ritter[1], W. Hiddemann, B. Wörmann, G. Schellong, Th. Büchner

Introduction

Recent improvements in cytogenetic technology have provided the means to identify chromosomal abnormalities in the majority of children with acute lymphoblastic leukemia (ALL) [4, 18]. In addition to numeric aberrations, specific nonrandom translocations have been described [4, 17]. Cytogenetic studies have also emphasized the clinical significance of chromosomal aberrations in addition to previously identified prognostic factors such as white blood count, sex and age [4, 11, 20]. The applicability of cytogenetic techniques, however, is limited by the laborious and time consuming processing of samples and therefore not suitable for routine clinical evaluations.

In contrast, flow cytometry (FCM) provides a rapid and reliable analysis of the cellular DNA content. Recent developments in flow systems and preparation techniques improved the accuracy of FCM measurements significantly. The present study was therefore carried out in 120 consecutive children with ALL from our institution in order to determine:

1. the incidence of DNA aneuploidies in childhood ALL as detected by flow cytometry,
2. the relation of DNA aneuploidies to patient characteristics such as sex and white blood cell count,
3. the relation of DNA aneuploidies to the phenotype of the leukemic cell,
4. the prognostic significance of DNA aneuploidies for therapeutic response and remission duration.

Patients, Treatment Protocols and Methods

From 120 consecutive patients with ALL admitted to our institution between July 1979 and March 1983 an evaluable DNA histogram was obtained in 112. All patients were entered into the BFM studies for the treatment of ALL [7, 8]; patients with B ALL received the BFM protocol for disseminated B-neoplasias [15]. In the study BFM 79/81 patients were stratified according to a risk index derived from a multiparameter Cox regression analysis in preceding studies. Children with a high risk index (> 2) received more intensive reinduction treatment than patients with a low risk index (0–2).

---

1 Univ.-Kinderklinik, Robert-Koch-Str. 31, 4400 Münster, FRG

Tumor Aneuploidy
Büchner et al.
© Springer-Verlag: Berlin Heidelberg 1985

In the study BFM 81/83 a modification of the risk index (risk factor) was applied taking into account the peripheral blast count, splenomegaly and hepatomegaly [12]. This risk factor was found to be of even higher prognostic significance than the risk index and is used for the treatment stratification in the present BFM study [8].

The diagnosis of ALL was based on the morphologic evaluation of Giemsa stained bone marrow smears and additional cytochemical (Periodic-acid-Schiff, acid phosphatase, peroxydase, alpha-naphtyl-acetate esterase) and biochemical (terminal transferase, TdT) [19] techniques. T-cell characteristics were identified by sheep erythrocyte rosette formation at 4°C [10], B-cell features by the presence of monoclonal surface immunoglobulin receptors [1].

For FCM analysis 0.5–1.0 ml of bone marrow and/or 1–5 ml of peripheral blood were aspirated into a syringe containing preservative free heparin (1000 U/ml) or EDTA (10 μg/ml) for anticoagulation. Bone marrow aspiration and blood specimens were subjected to Ficoll Hypaque gradient separation at $1000g$ for 20 minutes at room temperature. Cells were washed twice in Hanks medium, fixed in 96% ethanol and stored at 4°C. Mononuclear cells from sex-matched blood donors enriched by leukapheresis served for reference measurements and were prepared in the same manner. For FCM analysis the fixed cells were centrifuged again at $1000g$ for 5 minutes, the pellet was resuspended in 1 ml of a 0.5% pepsin-HCl solution for 1 minute and stained with Ethidium Bromide (10 mg/1000 ml Tris buffer pH 7.5) and Mithramycin (20 mg/1000 ml Tris buffer pH 7.5) in combination. For each histogram 20,000 to 50,000 cells were measured using an ICP 11 (Phywe AG) with a sheath flow glass chamber and a high accuracy 1024 channel multichannel analyzer. The determination of the DNA content was based on reference measurements with mononuclear cells from normal blood donors mixing patient and reference cells at ratios of 1:1 and 1:2. The ratio of the $G_{0/1}$ peaks of the patient sample and the reference cells provides the DNA index as introduced by Barlogie et al. [3]. The DNA index of normal diploid cells is 1.0. Based on previous studies defining a range of error of ± 3% for the determination of the DNA index, DNA indices ≤ 0.96 and ≥ 1.04 were considered as aneuploid.

Differences between two or more groups were evaluated with the chi-square test, with Yates' correction or by the Mann-Whitney-U-test. Life table analysis was performed according to Cutler and Ederer [6], possible differences between two or more groups were evaluated by the log-rank test.

Results

Applying reference measurements to all 112 evaluable patients aneuploid DNA stem lines were detected in 45 patients (40.2%). The DNA index ranged from 0.88 to 2.00 with a median of 1.18 (Table 1). Only one patient showed a hypodiploid DNA cell line, in another patient a tetraploid DNA stem line was found. Two different DNA stem lines were found in one boy with T-ALL. A significantly lower incidence of DNA aneuplodies was found in children with T-ALL (2/13; 15%) as compared to children with non-T/non-B ALL (28/62; 44%) (p < 0.05). Two of three children with B-ALL showed aneuploid DNA stem lines.

Table 1. Incidence of DNA aneuploidy in childhood ALL

	N	DNA Aneuploidy	DNA index
All children	112	45 (40.2%)	0.88–2.00 (median 1.18)
Non-T/non-B ALL	94[a]	41[a] (44%)	0.88–2.00 (median 1.20)
T-ALL (E$^+$)	13[a]	2[a] (15%)	1.07; 1.00 + 1.12[b]
B-ALL (s-Ig$^+$)	3	2 (67%)	1.06; 1.07

[a] $p < 0.05$; [b] two stem lines

Table 2. DNA aneuploidy in boys and girls with non-T/non-B ALL

	Diploid (n = 53)	Aneuploid (N = 41)
Boys (n = 46)	31[a]	15[a]
Girls (n = 48)	22[a]	26[a]

[a] $p < 0.05$

Table 3. Differences in presentation features of children with non-T/non-B ALL with (A) and without DNA aneuploidy

Prediagnostic history longer in A	($p < 0.001$)
WBC lower in A	($p < 0.01$)
Peripheral blast count lower in A	($p < 0.01$)
LDH lower in A	($p < 0.05$)

The DNA indices of blasts from patients with aneuploid T- or B-ALL were considerably lower (median: 1.07) than the DNA indices of blasts from patients with aneuploid non-T/non-B ALL (median: 1.20) (Table 1).

Significantly ($p < 0.05$) more girls (26/48; 54%) than boys (15/46; 32%) with non-T/non-B ALL revealed aneuploid DNA stem lines (Table 2). No differences in age were found between non-T/non-B ALL patients with (median: 5.7 years) or without (median: 5.2 years) DNA aneuploidies.

Further patient analysis revealed significant differences in the following characteristics: non-T/non-B ALL patients with DNA aneuploidies had a longer prediagnostic history ($p < 0.001$), a lower WBC and peripheral blast count ($p < 0.01$) and a lower level of lactat dehydrogenase in the serum than non-T/non-B ALL patients without detectable DNA aneuploidies (Table 3).

No differences were found between non-T/non-B ALL patients with or without DNA aneuploidies for hemoglobin-, uric acid-, alkaline phosphatase-, Ig G, A, M-levels, platelet count and hepatosplenomegaly. No differences were revealed either in the phenotypes of aneuploid vs non-aneuploid blasts according to the FAB classification, the cyto-chemical patterns and the TdT-activity.

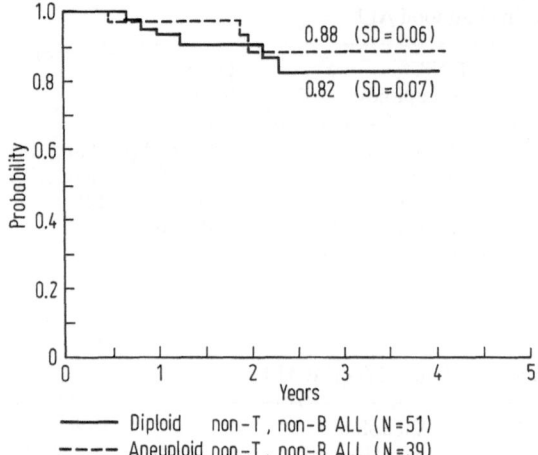

Fig. 1. Probability of continuous complete remission (life table method) for children with diploid and aneuploid non-T/non-B ALL under the conditions of the BFM protocols 79/81 and 81/83

Table 4. Treatment results in children with diploid and aneuploid non-T/non-B ALL (BFM protocols 79/81 and 81/83)

	Non-T/non-B ALL	
	Diploid	Aneuploid
Patients	53	41
Early death	1	1
Nonresponder	1	1 [a]
Complete remission	51 (96%)	39 (95%)
Death in remission	0	0
Relapse	6	3 [b]
CCR (life table) (median follow up: 34 months)	82±7%	88±6%

[a] DNA index at diagnosis 1.13; at relapse 1.13 + 1.40;
[b] DNA indices 1.06; 1.14; 1.15

Treatment results of children with non-T/non-B ALL are listed in Table 4. There were no differences between children with or without DNA aneuploidies for the rates of early deaths (1/1), non-responders (1/1) or complete remission (95%/96%). So far, no deaths in remission were observed.

After a median observation time of 34 months (range 5–50 months) six relapses occurred in patients without DNA aneuploidies as compared to three in patients with DNA aneuploidies. Relapse sites were not different between both groups. Based on life table analysis, no significant difference could so far be detected in the probability of continuous complete remission after 50 months being 88 ± 6% for patients with and 82 ± 7% for patients without DNA aneuploidies (Fig. 1).

Discussion

Recent improvements in the technology of flow systems and in cell preparation techniques have resulted in a high accuracy and reliability of flow cytometric analyses of the cellular DNA content. In addition, systemic reference measurements with mononuclear cells from sex matched normal blood donors have provided the means to detect aneuploid cell lines in a large proportion of patients with acute leukemias [9, 16]. In the present study DNA aneuploidies were identified in 45 of 112 children with ALL at diagnosis. The incidence of 40.2% DNA aneuploidies in childhood ALL exceeds previously reported data on FCM measurements in acute lymphoblastic leukemia, considerably [3, 13], and underlines the need for high quality technology and methodology of DNA measurements in acute leukemia. Taking into account that by FCM analysis only numeric chromosomal aberrations are detectable, these findings closely correspond to cytogenetic studies revealing numeric chromosomal abnormalities in 30% to 45% of all patients with evaluable metaphases [4, 18].

Biclonal leukemias were observed in one patient with T-ALL at diagnosis and in one girl at relapse when the reoccurrence of a previously identified hyperdiploid DNA cell clone was accompanied by the appearence of a second hyperdiploid DNA stem line. One patient showed a tetraploid DNA stem line at diagnosis. Look et al. [13] found an even higher incidence of biclonal ALL in children combining flow cytometry and karyotype analysis.

DNA aneuploidy as detected by FCM may correspond to the secondary chromosomal aberrations [21] while the primary chromosomal aberrations, such as structural abnormalities and specific translocations at or near cellular oncogenes [21], may not be detectable by FCM. While these primary chromosomal aberrations may be involved in the etiology of leukemia, secondary chromosomal aberrations may correspond with the biological behaviour of leukemia in a given patient.

According to this hypothesis significant differences between patients with non-T/non-B ALL with or without detectable DNA aneuploidies were found: patients with aneuploid blasts have a longer prediagnostic history, a lower WBC and peripheral blast count and a lower LDH in serum at diagnosis as compared to patients with diploid blasts. These parameters correlate with a somewhat smouldering form of ALL with less cell turnover. Similarly, cell kinetic studies have revealed a lower DNA S-phase index in peripheral blood cells in aneuploid non-T/non-B ALL [16].

Previous cytogenetic investigations revealed a correlation between chromosomal aberrations and clinically defined risk factors in childhood ALL such as T-cell phenotype, high WBC, sex and age. A significantly higher incidence of karyotype abnormalities in non-T/non-B ALL as compared to T-ALL was reported at the Third International Workshop on Chromosomes in Leukemia [4], a finding that could be confirmed in our study. Andreeff et al. [2] and Brodeur et al. [5] identified near-haploid ALL as a subtype with especially poor prognosis while Kaneko et al. [11] showed a longer remission duration in patients with chromosomal numbers over 50. In the present study no statistically significant difference in remission rate or remission duration was found between patients with aneuploid and diploid non-T/non-B ALL with a slight tendency towards longer remissions in patients with DNA aneuploidy. These findings are in contrast to results from the Memphis group [14] who found that patients with DNA indices above 1.15 had a significantly longer duration of complete remission on their standard risk protocol. The reason for this discrepancy is most likely the different therapeutic approach in both centers with striking differences in the percentage of long-term remissions. Hence, the intensified and prolonged BFM treatment with a low rate of relapses may compensate for the prognostic significance of DNA aneuploidy being obvious under less aggressive and successful therapy. These considerations are in accordance with the observation that under the conditions of an intensive and prolonged, risk-adapted induction therapy as in the BFM study most risk factors for relapse in childhood ALL such as T-cell phenotype [7], WBC [8] and sex [8] will lose their prognostic importance.

The results of the present study clearly indicate that FCM analysis provides a rapid and reliable evaluation of the cellular DNA content detecting DNA aneuploidies in a large proportion of children with ALL. Since aneuploidy is the most specific marker for malignancy small numbers of leukemic cells can be recognized objectively by flow cytometry when not yet identifiable by morphology for example in therapy-induced bone marrow hypoplasia or during complete remission thus inducing changes in the therapeutic strategy or detecting relapses at an early stage, respectively. Since the results from FCM analysis are available within hours, flow cytometry can easily be applied for routine clinical examinations also complementing and facilitating cytogenetic evaluation prior to therapy.

References

1. Aiuti F, Cerottini J, Coombs R, Cooper M, Dickler H, Froland S, Fudenberg H, Greaves M, Grey H, Kunkel H, Natvig J, Preud'homme J, Rabellino E, Ritts R, Rowe D, Seligman M, Siegal F, Terry W, Wybran J (1974) Special technical report: Identification, enumeration and isolation of B- and T-lymphocytes from human peripheral blood. Scan J Immunol 3:521–532
2. Andreeff M, Miller D, Steinherz P, Kempin S, Straus D, Clarkson B (1981) Hypodiploid acute lymphoblastic leukemia (ALL): a rare entity detected and monitored by flow cytometry. Proc AACR 22:44
3. Barlogie B, Hittelman W, Spitzer G, Hart JS, Trujillo JM, Smallwood L, Drewinko B (1977) Correlation of DNA distribution abnormalities with cytogenetic findings in human adult leukemia and lymphoma. Cancer Res 37:4400–4407
4. Bloomfield CD, Rowley JD, Goldmann AI, Lawler SD, Secker Walker LM, Mitelman F (1983) Chromosomal abnormalities and their clinical significance in acute lymphoblastic leukemia. Third International Workshop on Chromosomes in Leukemia. Cancer Res 43:868–873
5. Brodeur GM, Williams DL, Look AT, Bowman WR, Kalwinsky DK (1981) Near-haploid acute lymphoblastic leukemia: A unique subgroup with a poor prognosis? Blood 58:14–19
6. Cutler S, Ederer F (1958) Maximum utilsation of the lfe table method in analysing survival. J Chron Dis 4:699–713
7. Henze G, Langermann HJ, Fengler R, Brandeis M, Evers KG, Gadner H, Hinderfeld L, Jobke A, Kornhuber B, Lampert F, Lasson U, Ludwig R, Müller-Weihrich S, Neidhardt M, Nessler G, Niethammer D, Rister M, Ritter J, Schaaff A, Schellong G, Stollmann B, Treuner J, Wahlen W, Weinel P, Wehinger H, Riehm H (1982) Bericht über die Therapiestudie BFM 79/81 zur Behandlung der akuten lymphoblastischen Leukämie bei Kindern und Jugendlichen: Intensivierte Reinduktionstherapie für Patientengruppen mit unterschiedlichem Rezidivrisiko. Klin Pädiatr 194:195–203
8. Henze G, Langermann HJ, Riehm H (1982) Acute lymphoblastic leukemia therapy study BFM 81/83: Stratified treatment for patients with different risk for relapse estimated by the leukemic cell burden. 3rd Int. Symposium on Therapy of Acute Leukemias, Rome

9. Hiddemann W, Wörmann B, Ritter J, Henze G, Langermann HJ, Kaufmann U, Schellong G, Riehm H, Büchner T (1982) Diagnostik von Aneuploidien bei akuten Leukämien mittels Impulscytophotometrie (ICP): Häufigkeit und klinische Relevanz. Verh Dtsch Ges Inn Med 88:934–937

10. Jondal M, Holm G, Wigzell H (1972) Surface markers on human B- and T-lymphocytes. I. A large population of lymphocytes forming non immune rosettes with sheep red blood cells. J Exp Med 136:207–212

11. Kaneko Y, Rowley JD, Variakojis D, Chilcote RR, Check I, Sakurai M (1982) Correlation of karyotype with clinical features in acute lymphoblastic leukemia. Cancer Res 42:2918–2929

12. Langermann HJ, Henze G, Wulf M, Riehm H (1982) Mathematisches Modell zur Abschätzung der Blastenmasse der akuten lymphoblastischen Leukämie: Prognostische Bedeutung. Klin Pädiatr 194:209–213

13. Look AT, Melvin SL, Williams DL, Brodeur GM, Dahl GV, Kalwinsky DK, Murphy SB, Mauer AM (1982) Aneuploidy and percentage of S-phase cells determined by flow cytometry correlate with cell phenotype in childhood acute leukemia. Blood 60:959–967

14. Look AT, Melvin SL, Williams DL, Roberson PK, Bowman WP, Dahl GV, Murphy SB, Mauer AM (1982) Clinical and biological implications of flow cytometry determination of aneuploidy and pretreatment % S-phase of marrow blasts in childhood acute lymphoblastic leukemia (ALL). 6th Int. Symposium on Flow Cytometry, Elmau

15. Riehm H, Henze G, Langermann HJ, Odenwald E, Müller-Weihrich S (1983) Childhood B-type Non-Hodgkin's Lymphoma: Improved prognosis in two consecutive BFM studies. 13th Int. Congress of Chemotherapy, Vienna

16. Ritter J (1982) Akute Leukämien bei Kindern: Phänotyp und DNS-Gehalt der Blasten in Korrelation zum klinischen Erscheinungsbild und Krankheitsverlauf. Habilitationsschrift, Universität Münster

17. Roth D, Cimino M (1979) B-cell lymphoblastic leukemia with a 14q$^+$ chromosome abnormality. Blood 53:235–243

18. Rowley J (1978) The cytogenetics of acute leukaemia. Clin Haematol 7:385–406

19. Welte K, Ebener K, Hinderfeld L, Ritter J, Henze G, Kornhuber B (1981) Die Bedeutung der terminalen Deoxynucleotidyl-Transferase in der Diagnostik der akuten Leukämien des Kindes. Klin Pädiatr 193:165–171

20. Williams DL, Tsiatis A, Brodeur GM, Look AT, Melvin SL, Bowman WP, Kalwinsky DK, Rivera G, Dahl GV (1982) Prognostic importance of chromosome number in 136 untreated children with acute lymphoblastic leukemia. Blood 60:864–871

21. Yunis JJ (1983) The chromosomal basis of human neoplasia. Science 221:227–236

22. Zante J, Schumann J, Barlogie B, Göhde W, Büchner T (1976) New preparing and staining procedures for specific and rapid analysis of DNA distribution. In: Göhde W, Schumann J, Büchner T (eds) Pulse Cytophotometry II, European Press Medikon, Ghent 97–106

Incidence and Heterogeneity of DNA Aneuploidies in Solid Tumors and Acute Leukemias

W. Hiddemann[1], Th. Büchner, B. Wörmann, D. B. von Bassewitz, K. M. Müller, J. Ritter, J. Hauss, H. J. Kleinemeier, E. Grundmann

Introduction

At present, the most widely accepted theory about the pathogenesis of malignant tumors postulates a monoclonal or primarily unicellular malignant cell transformation [10, 12, 22]. Recent results from cytogenetic studies suggest that a specific non-random chromosomal aberration may be involved in this early step of cancer development [19, 34].

When the clinical diagnosis of a malignant tumor is made in a patient, however, a considerable heterogeneity within a single tumor cell population and also between the primary cancerous lesion and its metastases is found in most cancers expressing differences in growth rates, biochemical properties, potential to form metastases and karyotype abnormalities [1, 11, 13, 16, 21, 24, 33].

This intraneoplastic diversity is thought to be the result of a genetic instability of the malignant cell population [22] providing the emergence of heterogenous sub-populations with different characteristics and the occurrence of secondary chromosomal changes [19, 21, 26, 34].

Tumor heterogeneity has been the subject of several studies in the past, already, since its impact on the therapeutic management of patients with malignant diseases is recognized by many authors. Approaches that have been applied to evaluate the type and degree of heterogeneity within a tumor include the determination of hormon receptors [2, 29], in vitro growth characteristics such as plating efficacy and doubling time [16], biochemical assays [1], chromosome analyses [21], responsiveness to antineoplastic agents in vitro [8, 9, 17, 27] and measurements of the cellular DNA content by flow cytometry (FCM) [23, 28, 30].

In the present study we report on the results of FCM DNA analyses in 394 patients with acute leukemias and 155 patients with 3 different types of solid tumors which were performed to determine the overall incidence of DNA aneuploidies and the degree of tumor heterogeneity as reflected by the detection of multiple DNA stemlines.

1 Medizinische Klinik und Poliklinik der Universität, Albert-Schweitzer-Str. 33, D-4400 Münster, FRG

71

Tumor Aneuploidy
Büchner et al.
© Springer-Verlag: Berlin Heidelberg 1985

Material and Methods

Acute Leukemias

FCM DNA measurements were carried out on pretherapeutic bone marrow and/or blood samples from 394 patients with acute leukemia, comprising 119 adults with acute myeloblastic leukemia (AML), 52 adults with acute lymphoblastic leukemia (ALL), 43 children with AML and 180 children with ALL. Diagnoses were based on conventional cytologic, cytochemic, immunologic and biochemic (TdT) criteria.

Cell preparation consisted in the enrichment of the nucleated cells by density gradient separation over Hypaque Ficoll. The white cell layer was washed twice and fixed in 96% ethanol.

Prior to staining the cells were centrifuged again and the pellet was resuspended in a solution of ethidium-bromide and mithramycin at equimolar concentrations [5, 35].

Measurements were carried out on a modified ICP 11 with a sheath flow glass chamber [15]. For the detection of DNA aneuploidies sex matched mononuclear cells from normal blood donors were admixed as reference cells at two different concentrations [32]. The degree of DNA aneuploidies was expressed by the DNA index [6].

Solid tumors

The primary tumors of 155 patients with solid cancers were analyzed by FCM. 53 patients had large bowel carcinomas, 41 suffered from lung cancer and 61 from breast cancer. All diagnoses were established by conventional histologic and clinical features.

From every single tumor with the exception of breast cancers, samples from 5 up to 26 different sections (median 9) were analyzed by FCM, corresponding segments being evaluated by histology.

Cell preparation consisted in the mechanical dispersion of tumor samples, the subsequent filtering through nylon mashe gaze and fixation in 96% ethanol.

Staining and analysis were carried out as described before. Normal corresponding tissue from the same patient or mononuclear blood cells from normal blood donors served as references.

Results

Acute Leukemias

DNA aneuploidies were identified in 161 from 394 patients with acute leukemias (40%) with an almost identical frequency among the four major subgroups of adult AML (43%), adult ALL (40%), childhood AML (40%) and childhood ALL (40%). DNA indices ranged from 0.66 to 2.10 (Table 1).

Further analysis of immunologic subtypes in ALL revealed a significantly lower incidence of DNA aneuploidies in children with T-ALL (19%) as compared to non-T/non-B ALL (43%), a difference that seems to be the case for adult ALL as well with DNA aneuploidies in 40% of non-T/non-B ALL and in 24% of T-ALL.

Table 1. DNA aneuploidies in acute leukemias

		n	DNA aneuploidy	DNA index	
				Range	Median
Acute leukemias					
Children	ALL	180	72 (40%)	0.85–2.00	1.19
	AML	43	17 (40%)	1.05–2.00	1.12
Adults	ALL	52	21 (40%)	0.88–1.83	1.15
	AML	119	51 (43%)	0.66–2.10	1.09

Table 2. DNA aneuploidies in large bowel carcinomas, lung cancers and malignant breast tumors

	n	DNA aneuploidy	DNA index	
			Range	Median
Colon carcinomas	53	43 (81%)	1.06–3.64	1.75
Lung carcinomas	41	34 (83%)	0.75–3.40	1.70
Breast carcinomas	61	40 (66%)	1.07–2.34	1.77

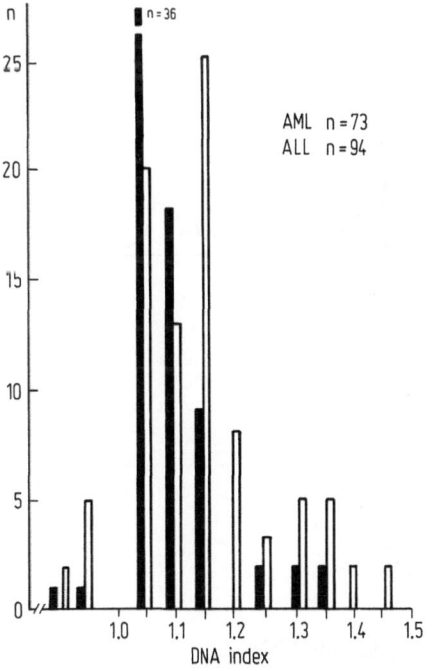

Fig. 1. Distribution of DNA indices in acute leukemias for ALL (open columns) and AML (solid columns)

The median DNA index was significantly lower in AML (median 1.09; 1.12) than in ALL (median 1.15; 1.19) both in adults and children (Fig. 1).

Multiple DNA stemlines were identified in 9 patients (6%) (Table 3), 5 adults with AML and 4 children with ALL.

In one patient with AML and with near haploid DNA aneuploidy (DNA index 0.66) the second clone was exactly double the value of the first one (DNA index 1.32) suggesting a multiplication or halfing of one originally common stemline. In the other 8 cases with more than one aneuploid DNA stemline no such relation was obvious. By morphology and cytochemistry alone a heterogenous leukemic cell population was suspected in 1 of the 9 cases, only.

Large Bowel Carcinomas

DNA aneuploidies were found in 43 out of 53 colon carcinomas (81%) with DNA indices ranging from 1.06 to 3.64, the median being 1.75 (Table 2).

More than one aneuploid DNA stemline was detected in 12 patients accounting for an incidence of 28% of aneuploid cases (Table 3, Fig. 2).

The DNA index of the second aneuploid cell line had exactly twice the value of the first DNA stemline in 7 of these 12 cases. In 5 patients no such relation was identified.

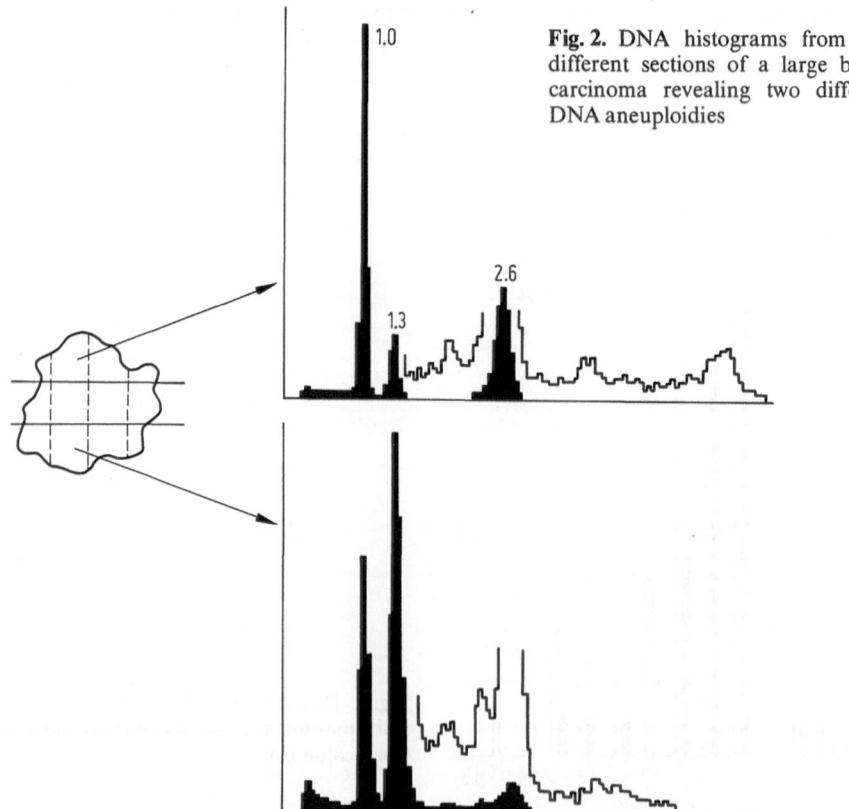

Fig. 2. DNA histograms from two different sections of a large bowel carcinoma revealing two different DNA aneuploidies

Table 3. DNA stemline heterogeneity in acute leukemias, large bowel carcinomas and lung cancers

	n	DNA aneuploidy	Multiple DNA stemlines
Acute leukemias	394	161 (40%)	9 (6%)
Large bowel carcinoma	53	43 (81%)	12 (28%)
Lung carcinomas	41	34 (83%)	15 (44%)

Histologic evaluation of the corresponding tumor sections revealed a uniform morphology of adeno-carcinomas in all cases.

Lung Carcinomas

34 from 41 lung carcinomas revealed DNA aneuploidies (83%) with a median DNA index of 1.70, range 0.75–3.40 (Table 2).

In 15 cases (44%) more than one aneuploid DNA stemline was identified, with twice the value for the second DNA index as compared with the first clone in 5 cases (Table 3). By histologic evaluation 30 carcinomas revealed a uniform morphology of

Table 4. Heterogeneity in lung cancer – a comparison of morphology and DNA aneuploidy

DNA stemline	Histology	
	Uniform	Heterogeneous
Monoclonal	22	4
Multiclonal	8	7

oat cell carcinoma (n = 12), squamous cell carcinoma (n = 13) or large cell carcinoma (n = 5). Heterogenous populations were identified in 11 cases.

7 of the 11 morphologically heterogenous lung cancers expressed more than one ancuploid DNA stemline, in 4 cases no heterogeneity of DNA aneuploidies was found. From 15 lung cancers revealing more than one aneuploid DNA stemline 8 were found to be heterogenous in their morphologic characteristics, as well, 7 expressed a uniform morphology (Table 4).

Breast Cancers

The overall incidence of DNA aneuploidies was 61% (40 from 61) with a median DNA index of 1.77, range 1.07–2.34 (Table 2).

Although no serial sections were investigated 4 tumors were found to express more than one abnormal DNA stemline (6.6%). Since this value is not considered representative for the incidence of heterogenous DNA stemlines in breast cancer the relation to histologic subtypes was not evaluated.

Discussion

The results of the present study on the overall incidence of DNA aneuploidies in three different types of solid tumors confirm the previously reported frequency of 70–90% DNA aneuploidies in most solid cancers analyzed so far [7, 14, 18, 23, 30]. In accordance with published data they also show an accumulation of abnormal DNA indices between 1.5 and 1.9 as a uniform finding independent on the tissue the tumor originated from [7, 18] (Fig. 3). In acute leukemias the median DNA indices were found to be significantly lower than in solid tumors as was the overall frequency of DNA aneuploidies. The incidence of 40% DNA aneuploidies found in the present study, however, exceeds previously published data reporting frequencies of DNA aneuploidies between 7 and 24%, considerably [4, 20].

Since artefacts due to an unbalanced staining of leukemic blasts and reference cells could be excluded by confirming DNA aneuploidies with different staining procedures, these results indicate the potential of high accuracy FCM measurements for the detection of abnormal DNA stemlines. They are also in good agreement with the frequency of numeric chromosome abnormalities detected by karyotype analysis, especially when taking into account that a considerable proportion of numeric chromosomal aberrations is not detected by cytogenetic techniques because of the lack of evaluable metaphases [3, 25, 31].

More than one abnormal DNA stemline was identified in 6% of acute leukemias, 28% of large bowel carcinomas and 44% of lung cancers. In malignant breast tumors multiple DNA aneuploidies were found in 7% of cases.

Since in breast cancer specimens only one area of each tumor was analyzed, in contrast to the evaluation of multiple different sites in large bowel and lung tumors, this low frequency of 7% multiple DNA aneuploidies is not considered representative.

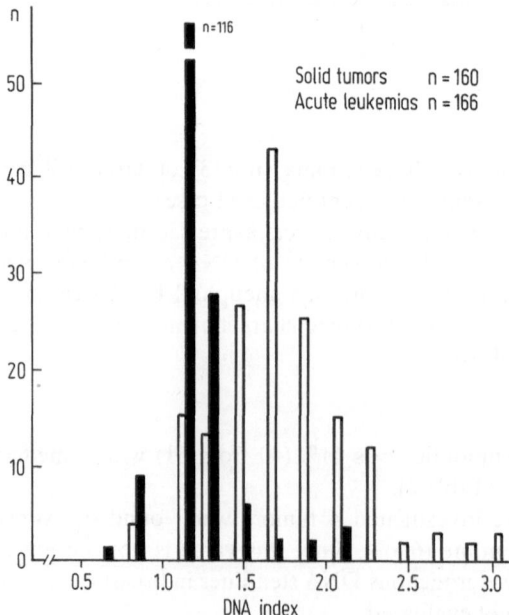

Fig. 3. Distribution of DNA indices in acute leukemias (solid columns) and solid tumors (open columns)

The difference in the frequency of multiple DNA aneuploidies between breast cancers versus lung and large bowel tumors emphasizes, however, the need for the analysis of multiple different parts of each tumor for the assessment of tumor heterogeneity.

The same argument may also explain the lower incidence of multiple DNA aneuploidies reported for lung and large bowel cancers in the literature [14, 23, 30].

In the present study 9 different specimens per tumor were analyzed by FCM and morphology on average (range 5–26). In nearly 50% of all samples only one abnormal DNA stemline was detected, indicating that the presence of multiple DNA aneuploidies is overlooked in approximately half the cases if only single specimens are examined.

In an attempt to elucidate the mechanisms leading to the occurrence of multiple DNA stemlines, the ratio between the different DNA indices within one tumor cell population was determined. 7 of 12 specimens from large bowel cancers and 5 of 15 lung cancer samples revealed a ratio of 1 : 2 for their DNA indices which may be interpreted as an indication for the emergence of the second aneuploid cell population from the first one by duplicating its DNA content i.e. by endomitosis. This interpretation has to be considered merely speculative since measurements of the cellular DNA content cannot provide the means to prove a clonal evolution which can be achieved only by high resolution cytogenetic analyses.

The detection of tumor heterogeneity and the underlying mechanisms leading to the development of heterogenous cancer cell populations gain increasing significance for the therapeutic approach to patients with malignancies. In the majority of most common solid tumors there is direct or indirect evidence of dissiminated disease at clinical diagnosis, already. Based on the findings discussed above intraneoplastic diversity must be assumed in a large proportion of cases. This might hamper any curative or adjuvant therapeutic modality in spite of successful local treatment because of the presence of subclones being resistant to the applied treatment.

Additional information about the nature of tumor heterogeneity and possible common features and relations between different subclones such as a uniform increase in the DNA content or the activation of macrophage reaction is needed to provide the rational for a more successful treatment of systemic malignant disease. In this context measurements of the cellular DNA content might help in detecting heterogenous tumor cell populations and in identifying common features among them.

References

1. Abeloff MD, Eggleston JC, Mendelsohn G, Ettinger DS, Baylin SB (1979) Changes in morphologic and biochemical characteristics of small cell carcinoma of the lung. Am J Med 66:757–764
2. Allegra JC, Barlock A, Huff KK, Lippman ME (1980) Changes in multiple or sequential estrogen receptor determinations in breast cancer. Cancer 45:792–794
3. Arthur DC, Bloomfield CD, Lindquist LL, Peterson BA, Nesbit ME (1981) Chromosome abnormalities in acute lymphoblastic leukemia (ALL): frequency and clinical implications. Proc Am Ass Cancer Res 22:345
4. Barlogie B, Büchner T, Hart JS, Ahearn MJ, Freireich EJ (1975) Aneuploidy as seen in the DNA-histogram of acute leukemia during the course of therapy. In: Haanen CAM, Hillen

HFP, Wessels JHC (eds) 1st Int Symp Pulsecytophotometry, European Press Medikon Ghent, p 299–305
5. Barlogie B, Spitzer G, Hart JS, Johnston DA, Büchner T, Schumann J, Drewinko B (1976) DNA histogram analysis of human hematopoietic cells. Blood 48:245–258
6. Barlogie B, Hittelman W, Spitzer G, Trujillo JH, Hart JS, Smallwood L, Drewinko B (1977) Correlation of DNA distribution abnormalities with cytogenetic findings in human adult leukemia and lymphoma. Cancer Res 37:4400–4407
7. Barlogie B, Raber MN, Schumann J, Johnston TS, Drewinko B, Swartzendruber DE, Göhde W, Andreeff M (1983) Flow cytometry in clinical cancer research. Cancer Res 43:3982–3997
8. Barranco SC, Ho DHW, Drewinko B, Romsdahl MM, Humphrey RM (1972) Differential sensitivities of human melanoma cells grown in vitro to arabinosylcytosine. Cancer Res 32:2733–2736
9. Biörklund A, Hakansson L, Stenstam B, Trope C, Akerman M (1980) On heterogeneity of Non-Hodgkin's lymphomas as regards sensitivity to cytostatic drugs. An in vitro study. Eur J Cancer 16:647–654
10. Cairns J (1981) The origin of human cancers. Nature 289:353–357
11. Dexter DL, Spremulli EN, Fligiel Z, Brabosa JA, Vogel R, VanVoorhees A, Calabresi P (1981) Hetergoneneity of cancer cells from a single human colon carcinoma. Am J Med 71:949–956
12. Fialkow PJ (1976) Clonal origin of human tumors. Biochem Biophys Acta 456:283–321
13. Fidler IJ (1978) Tumor heterogeneity and the biology of cancer invasion and metastasis. Cancer Res 38:2651–2660
14. Frankfurt OS, Slocum HK, Greco WR, Rustum YM (1982) Characterization of abnormalities in DNA content of human solid tumors. Proc Am Soc Clin Oncol 1:7
15. Göhde W, Schumann J, Büchner T, Otto F, Barlogie B (1979) Pulse cytophotometry: application in tumor cell biology and clinical oncology. In: Melamed MR, Mullaney PF, Mendelsohn ML (eds) Flow Cytometry and Sorting, John Wiley and Sons Inc, New York, 599–620
16. Gray JM, Pierce GB (1964) Relationship between growth rate and differentiation of melanoma in vivo. JNCI 32:1201–1210
17. Heppner GH, Dexter DL, DeNucci T, Miller FR, Calabresi P (1978) Heterogeneity in drug sensitivity among tumor cell subpopulations of a single mammary tumor. Cancer Res 38:3758–3763
18. Hiddemann W, Wörmann B, Ritter J, Kleinemeier HJ, von Bassewitz D, Roessner A, Müller KM, Büchner Th (1983) DNA aneuploidy – a highly specific marker for cancer detection. Proc Am Soc Clin Oncol 2:7
19. Levan G, Mitelman F (1981) The different origin of primary and secondary chromosome aberrations in cancer. In: Neth R, Gallo R, Graf H, Mannweiler K, Winkler K (eds) Modern Trends in Human Leukemia IV, Springer Verlag, Berlin Heidelberg New York, p 160–166
20. Look AT, Melvin SL, Williams DL, Brodeur GM, Dahl GV, Kalwinsky DK, Murphy SB, Mauer AM (1982) Aneuploidy and percentage of S-phase cells determined by flow cytometry correlate with cell phenotype in childhood acute leukemia. Blood 60:959–967
21. Mitelman F, Mark J, Levan G, Levan A (1972) Tumor etiology and chromosome pattern. Science 176:1340–1341
22. Nowell PC (1976) The clonal evolution of tumor cell populations. Science 194:23–28
23. Peterson SE, Bichel P, Lorentzen M (1979) Flow-cytometric demonstration of tumor cell subpopulations with different DNA content in human colo-rectal carcinoma. Eur J Cancer 15:383–386
24. Poste G, Fidler IJ (1980) The pathogenesis of cancer metastasis. Nature 283:139–146
25. Rowley JD (1978) The cytogenetics of acute leukemia. Clin Haematol 7:2–38
26. Sandberg A (1980) The chromosomes in human cancer and leukemia. Elsevier North-Holland, New York
27. Shapiro WR, Yung WA, Basler GA, Shapiro JR (1981) Heterogeneous response to chemotherapy of human gliomas grown in nude mice and as clones in vitro. Cancer Treat Rep [Suppl 2] 65:55–59

28. Slocum HK, Frankfurt O, Wake N, Pavelic ZP, Greco WR, Sandberg AA, Rustum YM (1982) Cellular heterogeneity of human solid tumors determined by flow cytometry, karyotyping and drug sensitivity in soft agar. Proc Am Ass Cancer Res 23:41

29. Sluyser M, van Nie R (1974) Estrogen receptor content and hormone-responsive growth of mouse mammary tumors. Cancer Res 34:3253–3257

30. Vindelov LL, Hansen HH, Christensen IJ, Spang-Thomsen N, Hirsch FR, Hansen M, Nissen MI (1980) Clonal heterogeneity of small-cell anaplastic carcinoma of the lung demonstrated by flow-cytometric DNA analysis. Cancer Res 40:4295–4300

31. Williams DL, Tsiatis A, Brodeur GM, Look AT, Melvin SL, Bowman WP, Kalwinsky DK, Rivera G, Dahl GV (1982) Prognostic importance of chromosome number in 136 untreated children with acute lymphoblastic leukemia. Blood 60:864–871

32. Wörmann B (1982) Intensivierte Induktionstherapie der akuten myeloischen Leukämie des Erwachsenen: Prognostische Bedeutung einiger zellulärer und zellkinetischer Parameter. Dissertation, Münster

33. Woodruff MFA, Ansell JD, Forbes GM, Gordon JC, Burton DI, Micklem HS (1982) Clonal interaction in tumors. Nature 299:822–824

34. Yunis JJ (1983) The chromosomal basis of human neoplasia. Science 221:227–236

35. Zante J, Schumann J, Barlogie B, Göhde W, Büchner T (1976) New preparating and staining procedures for specific and rapid analysis of DNA-distributions. In: Göhde W, Schumann J, Büchner T (eds) 2nd Int Symp Pulsecytophotometry, European Press Medikon, Ghent, p 97–106

Multiparameter Flow Cytometry for Determination of Ploidy, Proliferation and Differentiation in Acute Leukemia: Treatment Effects and Prognostic Value

M. Andreeff[1], A. Redner, S. Thongprasert, B. Eagle, P. Steinherz, D. Miller, M. R. Melamed

Abbreviations

ALL	= acute lymphoblastic leukemia	DAPI	= 4'-6-diamino 2 phenylindole
ANLL	= acute non-lymphoblastic leukemia	AMSA	= 4' (9-acridinamino) methansulfon-m-anisidide
FCM	= flow cytometry	CR	= complete remission
FITC	= fluorescein isothiocyanide	NR	= no response, failure

Total cellular DNA content measured by FCM has been studied by us [1–3] and others [4–8] for a number of years either alone or in conjunction with cellular RNA measurements [9] or with quantitative determinations of cell surface antigens [8, 10]. Abnormal DNA content, often found to be associated with abnormal chromosome numbers [11], is probably the most reliable "tumor marker" in clinical hematology, and in addition provides strong evidence for the clonality of hematopoietic tumors. Once a "DNA-stemline" is established, repeated samples are measured by FCM during induction, consolidation and maintenance therapy. The data provide important insights into the dynamics of cytoreduction and possibly differentiation of leukemia, especially in conjunction with conventional methods of cell counting and cell classification. The percentage of aneuploid cells detected by DNA FCM is often at variance with conventional blast cell counts and these discrepancies shed new light on the events involved in cytoreduction and cell differentiation. We also have used DNA measurements to monitor minimal residual disease and to detect relapse.

Measurements of DNA content also provide information regarding the proliferative activity of the cell population. Thymidine labelling index and S-phase fraction measured by FCM have previously been implicated in the prognosis of acute leukemias. In this report we have evaluated pretherapeutic cell kinetics, followed proliferative changes during induction therapy and correlated these results with clinical response.

Finally, the simultaneous measurement of DNA content and immunological properties as defined by polyclonal or monoclonal antibodies allows direct correlation of DNA aneuploidy, proliferation and differentiation. This provides a new tool for the understanding of the interactions between proliferation and differentiation and is also of major importance in understanding the effects of drugs on leukemic and non-leukemic cells.

---

1 Memorial Sloan-Kettering Cancer Center New York, New York 10021, USA.

Tumor Aneuploidy
Büchner et al.
© Springer-Verlag: Berlin Heidelberg 1985

Materials and Methods

Preparation of Cells

Bone marrow aspirates and Jamshidi biopsies were taken from the same posterior iliac crest after local anesthesia. Bone marrow (0.5–1 ml) was aspirated into heparinized syringes. Bone marrow biopsies were obtained from an immediately adjacent site and placed in 1–3 ml of medium TC 199 or Hank's balanced salt solution (HBSS). After complete mechanical dispersion the suspension was drained through a nylon filter to remove residual bone chips.

Bone marrow aspirates, biopsies and blood were subjected to Ficoll-Hypaque gradient separation. Spinal fluid (2 ml–4 ml) was spun down in its original test tube, the pellet was suspended in HBSS and stained as described below.

Simultaneous Staining for DNA and RNA

Aliquots (0.2 ml) of cell suspensions containing 0.2–0.4×10^6 cells in HBSS were mixed with 0.4 ml of 0.05 N HCl, 0.15 N NaCl and 0.1% Triton X-100 (v/v, Sigma Chemical Co., St. Louis, MO). After 30 seconds, 1.2 ml of a solution containing Na_2HPO_4 (0.2 M)-citric acid (0.1 M) buffer (pH 6.0), 1 mM EDTA-Na, 0.15 N NaCl and 6 µg/ml of acridine orange (AO) were added. Chromatographically purified AO obtained from Polysciences Inc. (Warrington, PA) was used. The pretreatment with Triton X-100 makes the cells permeable to the dye and at low pH nucleic acids remain insoluble. Subsequent staining with AO in the presence of the chelating agent EDTA and citric acid results in denaturation of all cellular RNA which then stains metachromatically red, while the native DNA intercalates the dye and stains orthochromatically green [9].

Simultaneous Staining for DNA and Cell Surface Antigens

Lyophilized Coulter (J5) clone reagent and other monoclonal antibodies were reconstituted by adding 0.5 ml of sterile distilled H_2O. Aliquots of 10 µl were stored at −80 °C, and were diluted with 390 µl of buffer (PBS and 0.2% BSA) for use. Staining was carried out as follows: cells were washed with buffer, 200 µl of 1 ° antibody was added and cells were incubated for 30 minutes at 4 °C and then washed twice with buffer. The supernatant was decanted and 200 µl of 2 ° antibody (Goat anti-mouse IgG (F(ab')$_2$, Cappel), was added (1:40 dilution) followed by incubation in an ice bath for 30 minutes. The cells were washed twice, fixed with 70% ethanol for 1 hour or overnight and washed again twice with HBSS. RNase was diluted with 1.12% Na citrate buffer, pH 8.4, to 500 U/ml (Rase, Worthington) and 250 µl was added, followed by incubation at 37 °C for 30 minutes. 250 µl of Propidium iodide was added and staining was done at room temperature for 30 minutes. The concentration of propidium iodide was 50 µg/ml in 1.12% Na citrate buffer [8, 10, modified].

Flow Cytometry Instrumentation and Multiparameter Data Processing

A computer-interfaced research cytofluorograf, model FC 200 (Ortho Instruments, Westwood, MA) was used to obtain simultaneous measurements of fluorescence of cells in two separate wavelength bands (F_{530} in a band from 515–575 mn; and

F > 600 in a band from 600–650 nm). In this instrument, the cells suspended in the dye solution are transported at rates of 200 per second. Fluorescence and light scatter signals were generated by each cell as it passed through the focus of a 488 nm argon-ion laser beam. The red fluorescence (F > 600) and green fluorescence (F530) emission for each cell was optically separated and measured by separate photomultipliers. 5000 cells were counted in each sample. Background fluorescence of the dye solution in which the cells were suspended was automatically subtracted. The pulse-width, i.e., the time the cell or nucleus takes to pass through the illuminating beam was recorded and used to distinguish single cells from cell doublets and other aggregates.

All the measured parameters were stored in computer memory (1220 Nova minicomputer, interfaced to the cytofluorograf). Software developed in our laboratory was used to analyze each sample separately. Statistical analysis was performed on the total and on subpopulations and includes total number of cells, mean, median, peak-values, standard deviation, co-efficient of variation, and skewness. Two parameter frequency histograms were obtained and displayed graphically by a Tektronix graphic terminal interfaced to the computer.

Cell Cycle Analysis

FCM derived DNA histograms can be analyzed in a number of ways to obtain cell cycle distribution, i.e. cells in $G_{0/1}$, S and $G_2 M$. In this study, we used a model proposed by us in 1973 [14], which is based on the assumption that the $G_{0/1}$ and $G_2 M$ peaks have a Gaussian distribution. Its shapes are determined by their left (for $G_{0/1}$) and right (for $G_2 M$) shoulders, respectively, which are flipped over to achieve symmetrical peaks, with cells between the two populations being in S-phase (Peak Symmetry Model). This model has the advantage of being applicable to both exponentially growing and perturbed cell populations. In order to validate this model, we compared it with the second order polynomial model proposed by Jett. Data from three histograms representative of nonperturbed low, intermediate and high proliferation were analyzed using both models (polynomial fit courtesy of Dr. J. Jett). Results are shown in Table 1. As expected, the results for low (S = 1%) and intermediate (S = 8%) proliferation were in excellent agreement, but Jetts' method calculated a significantly higher S phase for the highly proliferating cell population (40.6% vs. 29%), due to inclusion of a fair number of cells in S-phase which were

Table 1. Comparison of two methods for DNA histogram analysis

Bone marrow sample	GI %		S %		G_2M %	
	PSM	PFM	PSM	PFM	PSM	PFM
Low-S	98	97.8	1	1.7	1	0.5
Intermediate	89	90.2	8	8.7	3	1.1
High-S	60	54.0	29	40.6	11	5.4

PSM = Peak Symmetry Model
PFM = Polynomial Fit Model

under the $G_{0/1}$ and $G_2 M$ peaks. Similar distributions have recently been found with the BrDU/DNA technique, i.e. BrDU-incorporating cells were seen in the area of "G_1" and "G_2" peaks [15]. Since most human leukemias have low or intermediate number of S-phase cells (less than 15%), we feel that the application of our model to the investigation of leukemia cell kinetics is acceptable. During chemotherapy, the Polynomial Fit Model does not provide reliable cell cycle information because it may be difficult to identify distinct peaks, while the Peak Symmetry Model is still applicable, with certain limitations.

RNA Index

The RNA-Index is a standardized measurement of the RNA content of $G_{0/1}$ cells as compared with normal, unstimulated peripheral blood lymphocytes. It is calculated as: mean RNA content of $G_{0/1}$ cells of the sample $\times$ 10 divided by the median RNA content of control G_0 lymphocytes. It was shown to be a useful means to discriminate between acute lymphoblastic (ALL) and non-lymphoblastic leukemia (ANLL) (9), and has also been found to be one of the most important prognostic factors in adult ANLL [16] and ALL [17].

DNA Stemline

Abnormal DNA content, as measured by FCM, was established using the criteria of the "Convention on Nomenclature for DNA Cytometry" [12]. DNA distributions were extracted from DNA/RNA measurements using the acridine orange method, and from DNA/surface antigen measurements of propidium iodide/monoclonal antibody stained cell populations. The coefficient of variation (CV) was $2.9 \pm 0.9\%$ for control lymphocytes. Lymphocytes from healthy donors served as an external and internal standard for DNA stemline determination. They were measured separately, and then admixed to the leukemia samples, stained, measured and analyzed simultaneously. A widening of G_1, or a "shoulder", was not considered evidence for DNA aneuploidy, since a double peak was felt to be mandatory for the establishment of aneuploidy. For this purpose, controls, samples, and mixtures of controls and samples were measured at the high end of the DNA scale, in order to have more channels available for higher resolution. Some DNA distributions with shoul-

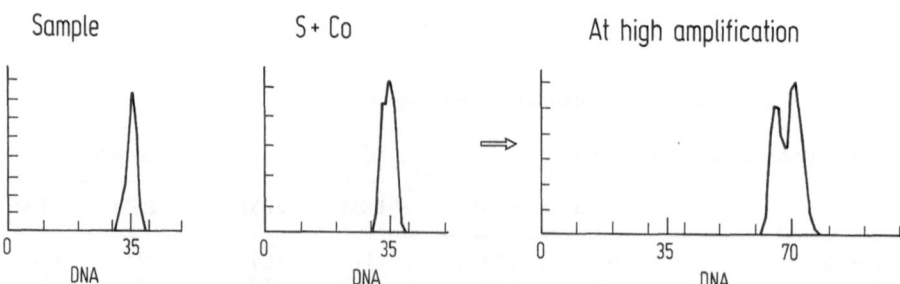

Fig. 1. DNA Histogram (AO green fluorescence) of leukemic bone marrow (sample). Addition of control lymphocytes adds to left shoulder of histogram (S+Co.). Measurement of same sample (S+Co) at high amplification reveals hyperdiploid DNA stemline (double peak), confirmed by cytogenetic analysis

Table 2. Induction chemotherapy protocols of acute non-lymphoblastic leukemia at Memorial Hospital

Protocol	ARA-C	TG	DNR or AMSA	
			DNR	AMSA
L-14M	25 mg/m² IV day 1 200 mg/m² IV Continuous infusion ×5days	100 mg/m² p.o. q 12 hrs×5 days	60 mg/m² IV×3 days	–
L-16	25 mg/m² IV day 1 200 mg/m² IV Continuous infusion ×5 days	100 mg/m² p.o. q 12 hrs×5 days	60 mg/m² IV×3 days	225 mg/m² IV ×3 days
L-16M	25 mg/m² IV day 1 160 mg/m² IV Continuous infusion ×5 days	100 mg/m² p.o. q 12 hrs×5 days	50 mg/m² IV×3 days	190 mg/m² IV ×3 days

ders at the low (normal) setting, which also included all cycling cells, could be resolved into double peaks at the high setting (Fig. 1). Thus, as low as a 5% difference in DNA content could be resolved and this was denoted as 2.1 c (2.0 c being diploid).

Treatment Protocols

The treatment protocols for adult ANLL patients are shown in Table 2. They employ the anthracycline Daunomycin (DNR) or the acridine derivative AMSA in combination with Cytosine-Arabinoside (ARA-C) and 6-Thioguanine (TG). The protocols used for ALL are described by Clarkson and co-workers [26].

Results

DNA Stemlines in Untreated ALL

One hundred thirty four pediatric and thirty-nine adult ALL patient samples were analyzed prior to treatment. DNA aneuploidy was found in 52/134 = 38.8% of pediatric and in 8/39 = 20.5% of adult cases. The distribution of DNA stemlines ranged from near haploid to hypertetraploid with the highest frequency at 2.4C, i.e. 20% higher DNA content compared to normal.

DNA stemlines in Untreated ANLL

Thirty-five pediatric and one hundred and twenty adult ANLL patient samples were analyzed prior to treatment. DNA aneuploidy was found in 5/35 = 14.3% of children and in 25/120 = 20.8% of adult cases. The distribution of DNA stemlines ranged from markedly hypodiploid (1.4c) to hyperdiploid (2.1c) in children and to hypertetraploid (4.3c) in adults, with the highest frequency at 2.2c, i.e. a 10% higher DNA content compared to normal.

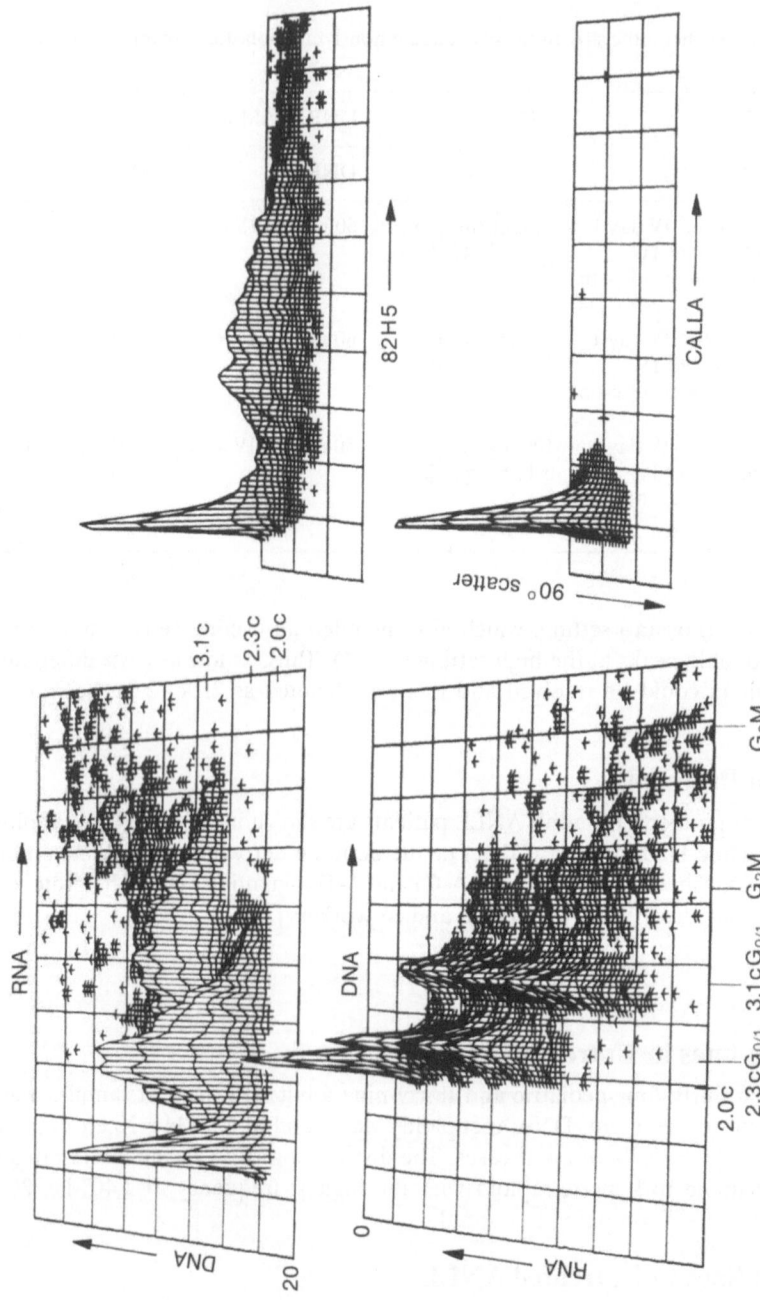

Fig. 2. Pediatric AML in relapse. Hyperdiploid (2.3c) and hypertriploid (3.1c) cell populations can be identified with their proliferative compartments. RNA content is high, characteristic for AML. The two left hand panels display the same DNA/RNA measurement from different axes: RNA (top) and DNA (bottom). Right hand panels show measurements of 90° scatter signals and monoclonal antibodies: positivity for the early myeloid antigen 82H5, negativity for the lymphoid antigen CALLA

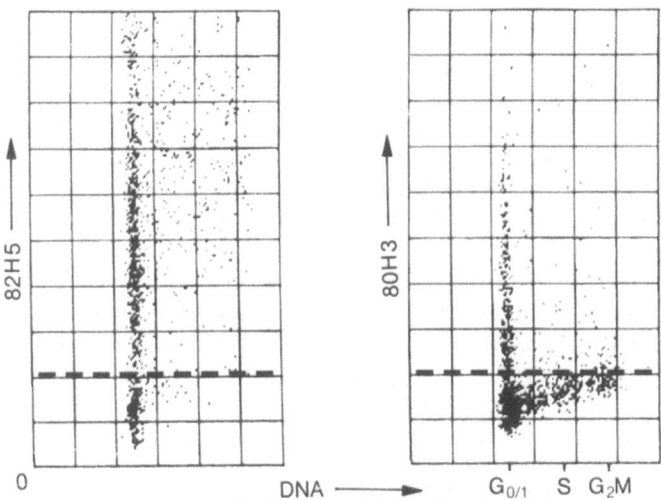

Fig. 3. Correlated FCM measurements of DNA (x-axis) and early (82H5) and late (80H3) myeloid differentiation antigens in the human leukemia cell line HL-60. Cells in $G_{0/1}$, S and G_2M are positive for the early myeloid antigen (left). The late myeloid antigen (right) is expressed by some $G_{0/1}$ cells but not expressed by S and G_2M cells. The dotted line separates negative and positive cells

Multiple DNA Stemlines

A small number of acute leukemias initially or at relapse presented with more than one stemline: diploid and aneuploid or different aneuploid cell populations may co-exist. An example is given in Fig. 2: this pediatric AML has diploid (2.0c), hyperdiploid (2.3c) and hypertriploid (3.1c) cells. Proliferating cells for each popula tion can be identified and the RNA index for both aneuploid myeloblastic populations is characteristically high. The maximal number of DNA stemlines observed in acute leukemia was four, ranging from diploid to hypertetraploid (pediatric ALL).

Simultaneous Measurement of DNA and Surface Antigens in Leukemia

Flow cytometric measurement of cell surface antigens has become the standard method for the evaluation of differentiation antigen expression. An example is given in Fig. 2 (right panel). It shows positivity for the early myeloid differentiation antigen 82H5 and negativity for the common ALL antigen. Simultaneous staining for DNA allows us to answer two important questions:
1. is the antigen expression related to the cycle, and
2. is the antigen expressed solely on aneuploid leukemic cells or also on diploid cells that may or may not be leukemic.

In our evaluation of monoclonal antibodies that are of potential use for studying differentiation in ANLL, we measured early and late myeloid antigens defined by the monoclonal antibodies 82H5 and 80H3 (gift of Dr. Mannoni) [18]. Exponentially growing HL-60 human promyelocytic cells were used for these studies [19]. Figure 3 shows that 85% of HL-60 cells express the early antigen defined by 82H5 at a very high level and that essentially all proliferating cells are positive for 82H5. HL-60 cells undergo some spontaneous differentiation and the late antigen 80H3 is expressed on 27% of these cells. Of interest is the negativity of virtually all proliferating cells for this antigen (80H3). This experiment clearly identifies the cell cycle specificity of these early and late myeloid differentiation antigens.

A panel of monoclonal antibodies is routinely used to characterize differentiation in ALL and an example is given in Fig. 4. DNA/RNA FCM of bone marrow biopsy and blood cells shows the existence of markedly hypodiploid (1.1c) and slightly hyperdiploid (2.1c) cells. Double-staining for DNA and Ia-like antigen, common ALL antigen, and the pre-B cell antigens BL1 and BL2 shows the presence of these differentiation antigens on both hypodiploid and hyperdiploid cells including proliferating cells (1.1c/2.1c Ia$^+$, CALLA$^+$, BL1$^+$, BL2$^+$). The presence of these antigens on the near-diploid cells (2.1c) is additional proof of their leukemic nature.

Proliferation of Acute Leukemia Prior to Treatment

Our initial series of *adult* leukemias showed no differences between ALL and ANLL regarding the S-phase compartment [ALL: S = 5.1 ± 4.1% (n = 26); ANLL: S = 4.1 ± 2.3% (n = 46)]. These data were obtained from bone marrow aspiration material [9]. A subsequent study compared aspiration and biopsy material in ANLL and found a biopsy S-phase of 11.7%, as compared to an aspiration S-phase of 7.31% (n = 66) [6]. The same study did not demonstrate a correlation of pretreatment S-phase with therapeutic response.

The present study of cell kinetic changes in adult ANLL during induction chemotherapy, based on bone marrow biopsy material, showed an S-phase of 6.93% (Table 3) and an RNA-Index of 16.32 prior to treatment (Table 4). DNA/RNA FCM of bone marrow aspirates in 111 patients with *pediatric* ALL showed an overall S phase of 6.7 ± 5.8%, with a significant difference between FAB L1 (5.8 ± 4.6) and L2 (12.4 ± 8.9) and diploid (S = 4.1%) and aneuploid (S = 8.3%) L1 cases (p = 0.001) [21]. Likewise, RNA content was significantly higher in aneuploid (RNA-Index 12.4 ± 3.7) than in diploid (10.7 ± 2.4) ALL (p = 0.006).

Comparison of Blast Number and Number of Aneuploid Cells in Untreated Leukemia

In untreated leukemia there was a difference between the number of blast cells counted on marrow smears and the number of aneuploid cells measured by FCM in aspirate and biopsy material. In the patients with aneuploid stemlines aspirates and biopsies were obtained simultaneously. In aspirates the number of aneuploid cells exceeded the blast number by 28%, while the aneuploid cells in biopsies exceeded the blast percentage by 26%. In some cases a 2-fold higher number of aneuploid cells was measured (Table 6, Fig. 10).

Cell Kinetic and RNA Content Changes During Induction Therapy of ANLL

The overall response rate was 18/28 = 65% achieved complete remissions (CR) and 10/28 = 35% failed (NR). CR patients had the same age and sex ratio as NR pa-

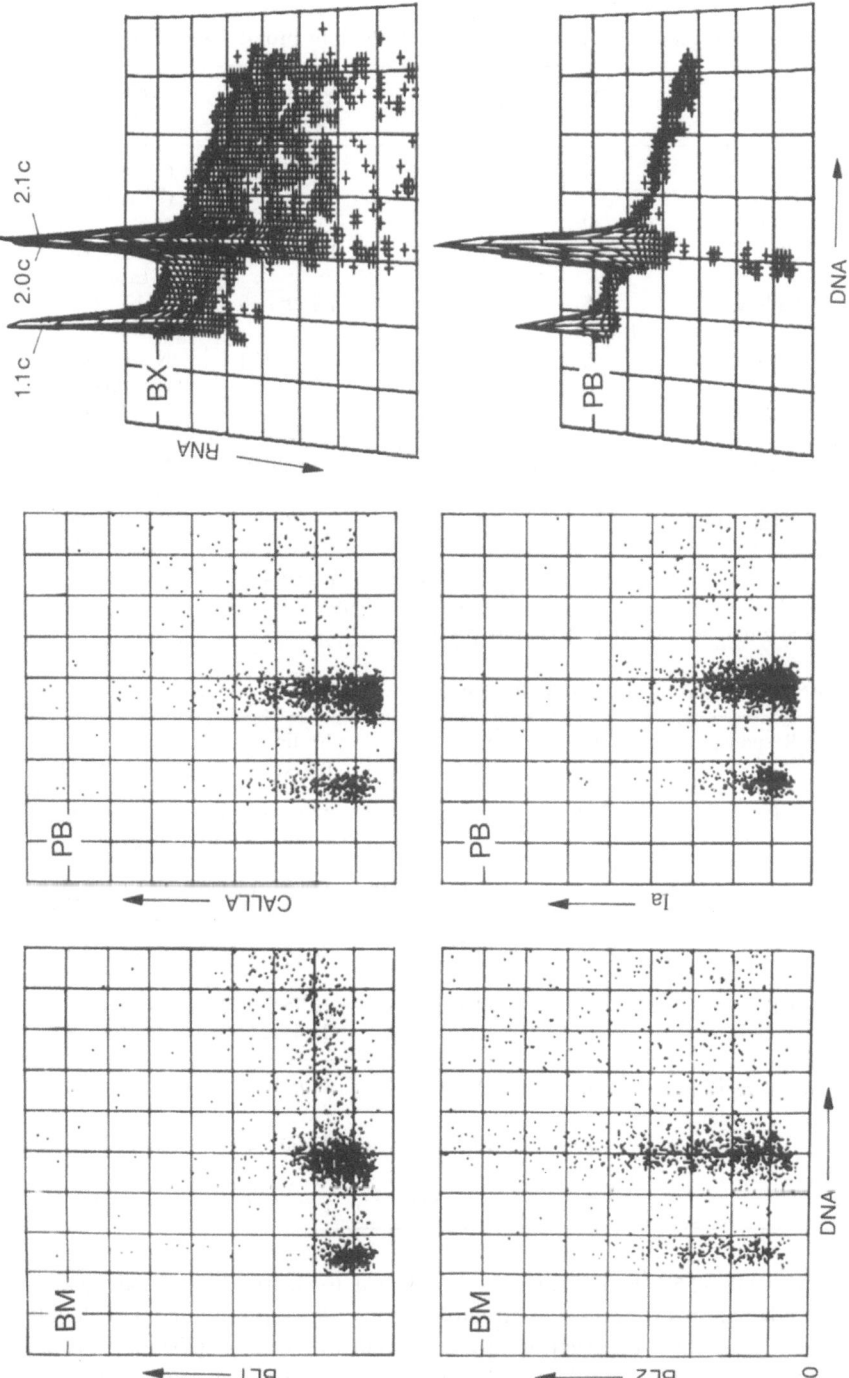

Fig. 4. Right panel shows correlated DNA/RNA histograms of ALL biopsy (Bx) and blood (PB) cells. Hypo-(1.1c) and hyperdiploid (2.1c) cells can be identified. The left hand panels show correlated measurements of DNA (x-axis) and different antigens (y-axis). Both hypo- and hyperdiploid cells are positive for Ia. CALLA, BL1 and BL2

Table 3. S-phase during and after induction therapy of ANLL by protocol and response

Protocol	Response	n	pre-Rx S±S.D.	Nadir S±S.D.	Recovery S±S.D.
L-14M	CR	4	4.5 ±2.65	0.25±0.5	6 ±3.61 (3 cases)
	Failure	0			
L-16	CR	12	8.25±3.82	1.5 ±0.83	7.11±4.37 (9 cases)
	Failure	8	6.50±3.16	1.81±1.69	10.17±8.04 (6 cases)
L-16M	CR	2	8 ±8.49	2.5 ±0.71	No data on recovery
	Failure	2	4.5 ±3.54	6.0	8.5 ±4.95
	Total CR	18	7.39±4.18	1.33±0.99	6.83±4.06 (12 cases)
	Total failure	10	6.1 ±3.14	2.65±2.3	9.75±7.09 (8 cases)
			p=0.404	p=0.04	p=0.256

Decrease in S: difference between CR/failure p=0.095

Table 4. RNA-index during and after induction therapy of ANLL by protocol and response

Protocol	Response	n	pre-Rx RNA±S.D.	Nadir RNA±S.D.	Recovery RNA±S.D.
L-14M	CR	4	16.7 ±5.19	14.55±2.66	16.5 ±3.80 (3 cases)
	Failure	0			
L-16	CR	12	17.09±2.49	11.49±1.95	12.33±3.69 (9 cases)
	Failure	8	14.78±3.72	12.14±2.31	15.82±4.50 (6 cases)
L-16M	CR	2	12.4 ±0.99	10.75±0.92	No data
	Failure	2	16.75±4.17	11.45±1.20	14.10±2.97
	Total CR	18	16.5 ±3.3	12.08±2.37	13.38±4.01 (12 cases)
	Total failure	10	16.0 ±3.0	12.00±2.1	15.38±4.04 (8 cases)
			p=0.343	p=0.922	p=0.287

tients, but 10/18=56% had Auer-rods, compared to 3/10=33% in NR patients (Table 5). NR patients were more frequently positive for terminal deoxynucleotidyl transferase (TdT). These observations are consistent with data published previously for a much larger patient population [20]. Typical changes in *cell kinetics* and RNA content in bone marrow biopsies during induction therapy are shown in Fig. 5: on day 0, cells are characterized by high RNA content (RNA-index: 14.6) and high proliferation (S=10%). On day 7, following 5 days of chemotherapy, a decrease in RNA

Table 5. Treatment of ANLL in adults; clinical characteristics

	Responders (CR)	Non-responders (NR)
Number of cases	18	10
Median age (years) (range)	45 (17–68)	47.5 (23–69)
Sex F:M	13:5	7:3
Auer rods positive	10/18	3/10
TdT positive	1/13[a]	4/6[a]
Morphological classification (FAB)		
M1	2	3
M2	6	1
M3	3	0
M4	3	3
M5	4	3

[a] Number of cases studied

index to 10.6 and proliferation to S=2% is characteristic and on day 13 RNA and proliferation recover (RNA-Index = 10.7, S=8%). The dynamics of these changes can be seen on Fig. 6 for responders and on Fig. 7 for non-responders on the L-16 protocol. For all CR patients, S-phase decreased to $1.33 \pm 0.99\%$ and to 2.65 ± 2.3 for failures. This difference in nadir S-phase between CR and failure patients is significant on the $p=0.04$ level (Table 3). There was no difference between CR and failures regarding their pretreatment ($p=0.404$) or recovery ($p=0.256$) S-phases and the absolute decrease in S was also not significantly different between the two groups ($p=0.095$).

The RNA-Index decreased significantly from 16.3 to 12.0 during induction therapy and recovered to 14. There was no significant difference between CR and failures for each time point analyzed.

Data show that chemotherapy with DAT or AAT exhibits a cell cycle specific effect that leads to a significant decrease of proliferating cells and also to a marked decrease of the RNA content of $G_{0/1}$ cells. Pretreatment S-phase does not appear to be a significant predictor of response, but in patients who achieve CR, the nadir S phase is significantly lower than in those who fail, indicating a more complete eradication of proliferating cells and/or block of entry into the S-phase. The kinetics of recovery does not seem to be a measure of response [22].

Changes in the Number of Blasts and Aneuploid Cells During Induction Therapy of ALL

A systematic comparison was carried out to better determine the number of leukemic cells during induction therapy. Though differences between blast number and the number of aneuploid cells were seen in most cases prior to treatment, these differences were amplified during induction therapy [23]. Figure 8 shows histograms from a patient with ALL, whose marrow contained 76.5% blasts and 89.1% mark-

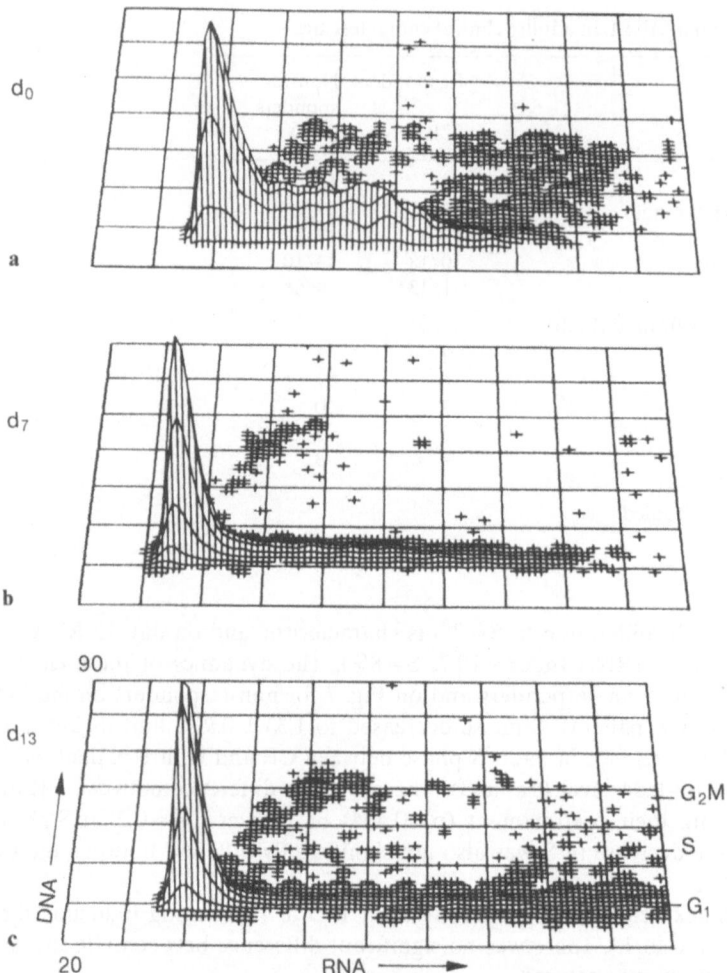

Fig. 5a–c. Change of DNA/RNA histograms during induction therapy for ANLL with DAT/ AAT. **a.** *Day 0:* Cells are characterized by high RNA content (RNA-Index: 14.6) and high proliferation (S-phase: 10%); **b.** *Day 7:* Decrease of RNA content (RNA-Index: 10.6) and proliferation (S-Phase: 2%); **c.** *Day 13:* Recovery of proliferation (S-Phase: 8%) and slight increase in RNA content (RNA-Index: 10.7)

edly hyperdiploid (2.5c) cells prior to treatment. On day 14 of induction therapy, blasts were not evaluable in the aspirate or in the markedly hypocellular biopsy, but the latter contained 36.4% hyperdiploid cells by FCM analysis. In another patient (Fig. 9) with known hyperdiploid DNA stemline (2.4c), 24.8% aneuploid cells were seen in the aspirate, and 52.8% in the biopsy, with high proliferation but in the absence of morphologically identifiable blast cells. On average during induction therapy, the aspirate had 3.05 times as many aneuploid cells as blast cells and the biopsy 6.15 times as many. The highest ratio was 7.63 for aspirate and 27.5 for biopsy ma-

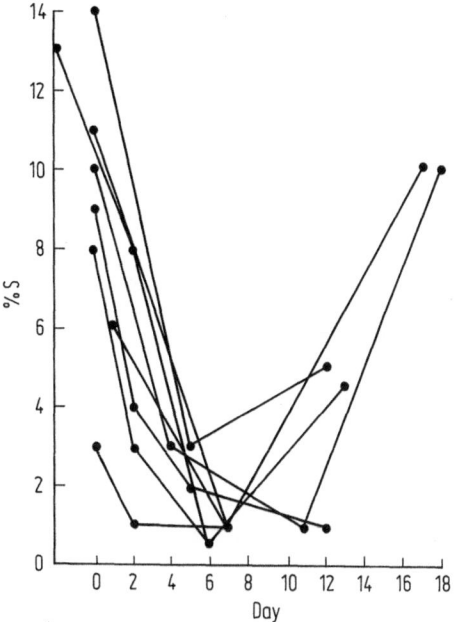

Fig. 6. Change in proliferation (FCM S-Phase) in bone marrow biopsies of ANLL in responding (CR) patients treated according to the L-16 protocol

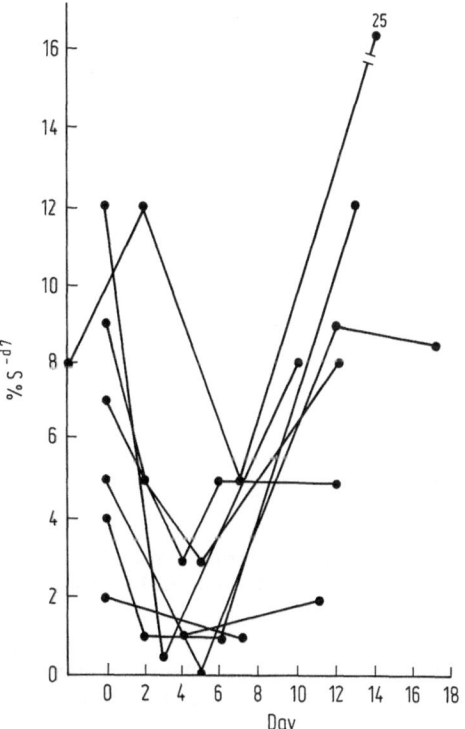

Fig. 7. Change of FCM S-Phase during induction therapy with L-16 protocol: decrease and recovery of proliferation in bone marrow in non-responding (NR) patients

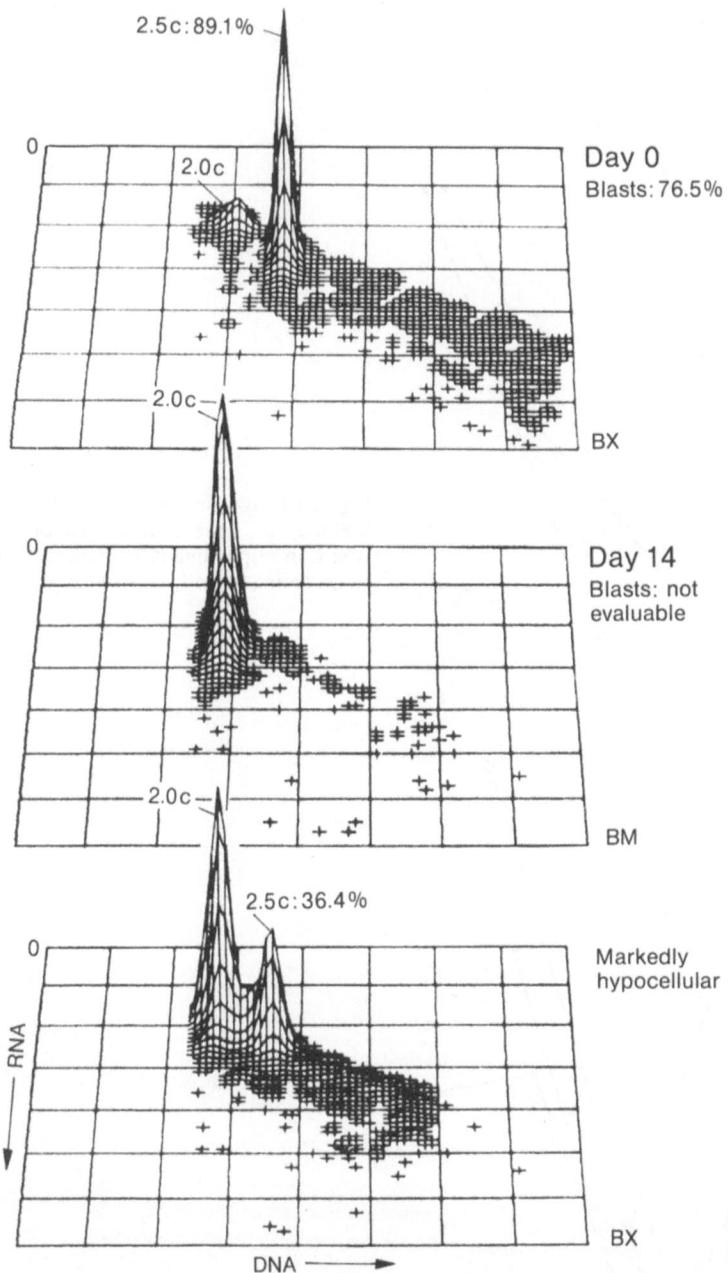

Fig. 8. Monitoring of aneuploid leukemic cells (ALL) in bone marrow biopsies prior to treatment and on day 14. Note the presence of 36.4% hyperdiploid cells in the d14 biopsy (Bx) with no evaluable blasts in the marrow aspirate (BM)

Fig. 9. Comparison of aneuploid cells and blasts on d14 of therapy. Prior to therapy, a hyperdiploid DNA-stemline (2.4c) was identified. On day 14 no blasts, but 24.8% (BM=aspirate) and 52.8% (Bx=biopsy) hyperdiploid cells are present

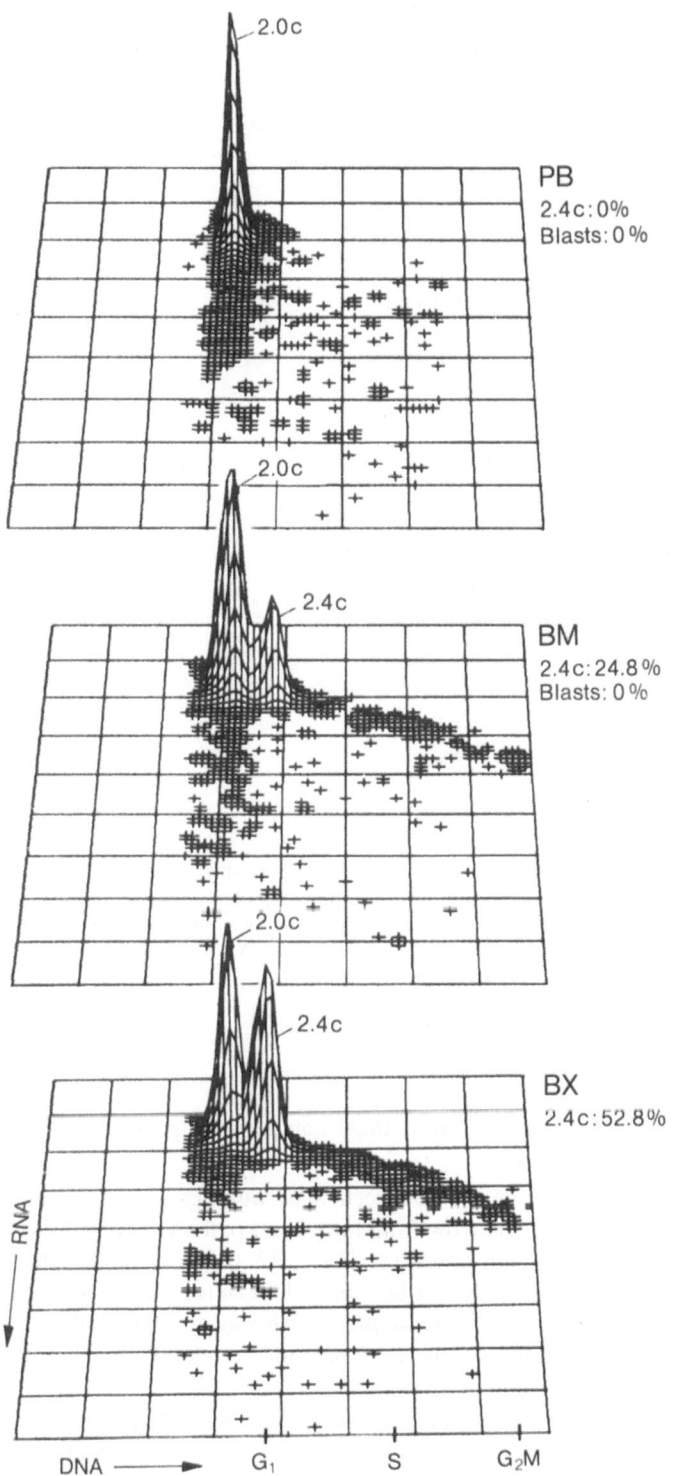

Table 6. Comparison of the percentage of blast cells and the percentage of DNA aneuploid cells in bone marrow aspirates and biopsies before and during induction chemotherapy

		Mean	Range
Pretreatment	% aneuploid cells in bone marrow aspirate/% blasts	1.28	0.94–2.07 n = 13
	% aneuploid cells in bone marrow biopsy/% blasts	1.26	0.94–2.02 n = 13
During induction	% aneuploid cells in bone marrow aspirate/% blasts	3.05	1.02–7.63 n = 11
	% aneuploid cells in bone marrow biopsy/% blasts	6.15	1.04–27.5 n = 12

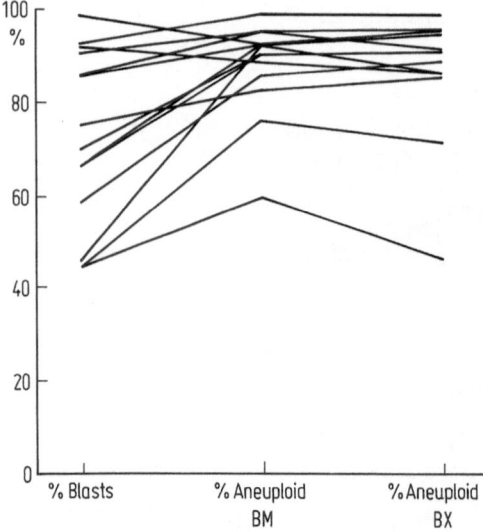

Fig. 10. Comparison of marrow blasts and aneuploid cells in acute leukemia prior to therapy (BM = aspirate, Bx = biopsy). Each line represents one patient

terial (Table 6) and Figs. 10 and 11 show these comparisons prior to and during treatment for all patients studied.

Clinically important "discrepancies" between blast number and aneuploid cell number were found, when bone marrow from individual patients were studied serially. The results of 3 typical patients are shown in Fig. 12. In these patients, the blast number dropped to less than 3%, but 5 to 18% aneuploid cells could be identified at any given time point, suggesting residual disease which was not detected by examination of cell morphology. Subsequently these patients "relapsed", with high blast numbers and increase in aneuploid cell number.

It must be emphasized that interactions of drugs with the DNA of leukemic or normal cells, or with the binding of the fluorochromes used (acridine orange and propidium iodide) to the DNA played only a minor role. Prednisone, vincristine, cy-

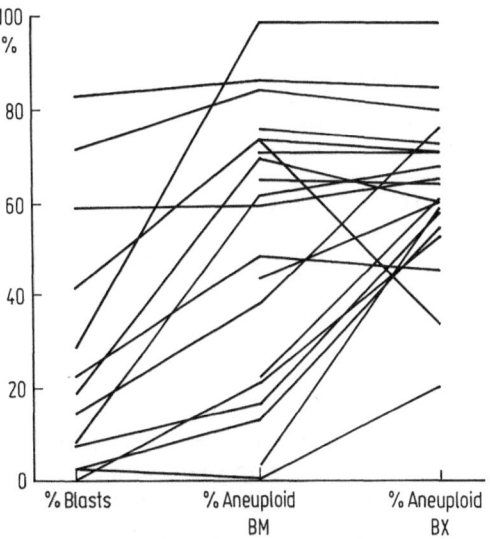

Fig. 11. Comparison of marrow blasts and aneuploid cells in acute leukemia during induction therapy. Note the differences between blast percentage and percentage of aneuploid cells

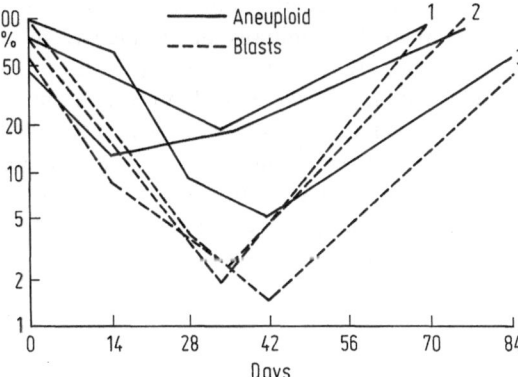

Fig. 12. Comparison of percentages of DNA aneuploid and blast cells in acute leukemia during and following induction therapy. Note: percent scale is logarithmic. Persistence of more than 5% aneuploid cells is followed by overt clinical relapse

toxan and most other drugs that were used in the treatment of most patients did not interfere with FCM measurements. This could be different in some patients who were treated with intercalating drugs such as daunomycin (DNR), but even high DNR or m-AMSA levels rarely shifted the $G_{0/1}$ peak in comparison to normal lymphocytes. In such cases, other dyes (DAPI, Hoechst) were employed to determine the correct DNA stemline, but as a matter of caution, DNA aneuploidy was only diagnosed in untreated patients and then followed sequentially.

Monitoring of Submicroscopic Levels of Aneuploid Cells During Maintenance Therapy

Once DNA aneuploidy was established, patients were followed during induction, consolidation and maintenance therapy. A group of 93 aneuploid pediatric ALL pa-

Table 7. Monitoring of aneuploid pediatric ALL by FCM

No. of follow-up studies (93 patients)		Length of follow-up studies (months)	DNA stemline	
			(n = 9) hypo	(n = 84) hyper
Mean	5.81	15.85	1.40 C	2.44 C
SD	5.16	16.99	0.30 C	0.38 C
SE	0.54	1.76	0.101 C	0.041 C
Range	1–27	0.4–69.1	1.1–1.8 C	2.1–4.1 C

tients was followed with multiple blood and bone marrow studies to detect early relapse, which was expected to coincide with the reappearance of aneuploid cells (Table 7). The follow-up period ranged from 2 weeks to 5.75 years, with an average period of 15.9 months. Between one and 27 follow-up samples were studied, with an average of 5.8 samples per patient. The average interval between samples was therefore 2.7 months. Nine patients had hypo- and 84 patients had hyperdiploid DNA stemlines. Four patients relapsed clinically. In two patients marrow samples obtained 37 and 60 days prior to clinical relapse showed 58% and 21% aneuploid cells, respectively. The stemlines were identical to those initially determined (2.4c and 2.3c). In the other two patients, marrow aspirates 3 and 6 months prior to relapse did not show evidence of aneuploid cells. No biopsies were available at that time.

These results can be improved, when fewer proliferating cells generate a lower "background" for hyperdiploid $G_{0/1}$ cells. This situation is found in the cerebrospinal fluid (CSF). The results of FCM of CSF to detect initial central nervous system leukemia are reported elsewhere (24). In one patient, the aneuploid leukemic clone reappeared six months prior to clinical relapse (Fig. 13).

Discussion

In this report, we describe methods for the quantitative analysis of DNA content, RNA content and surface antigen expression of individual cells by multiparameter FCM, their application to the study of acute leukemia, their prognostic relevance and their use for monitoring patients during chemotherapy.

DNA-Aneuploidy

Our data on DNA-ploidy determinations by FCM were obtained with a slightly modified technique that allowed us to better identify small differences in DNA content. Differences of 5% can now be determined with a high degree of certainty. This is important in the study of acute leukemia which show mostly minimal DNA deviations from normal. The most frequent DNA stemline in ANLL was 2.2c, and 2.4c in ALL.

The aneuploidy rate was 38.8% in pediatric *ALL* and only 20.5% in adult ALL. In the light of recent reports on the favorable prognostic importance of hyperdiploid DNA stemlines in ALL [28], the lower frequency in adults may account in part for

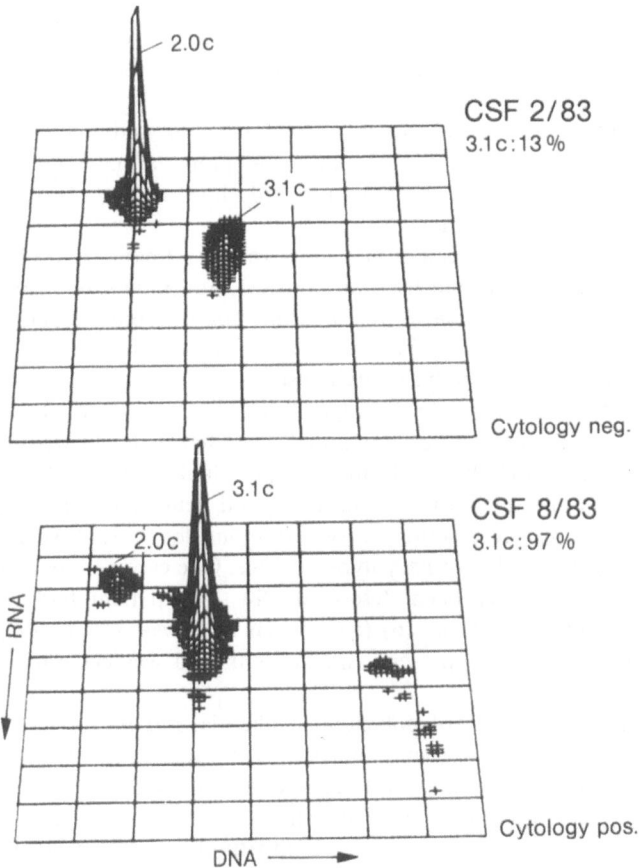

Fig. 13. Detection of aneuploid cells in the spinal fluid of a patient with ALL six months prior to clinical CNS relapse

their poorer prognosis. This however has to be studied prospectively. The poorer prognosis of DNA-diploid ALL seems plausible since patients whose marrow cells have pseudodiploid karyotypes by cytogenetic analysis fall into this group. Patients with pseudodiploidy are now considered to have a poor prognosis [34]. Another group with ploidy-defined poor prognosis have near-haploid DNA stemlines [25], and this group will be reported in detail elsewhere (paper in preparation).

The lower number of patients with detected DNA aneuploidy in *ANLL* may in part be related to the fact that this type of leukemia is not associated with a high degree of aneuploidy. Differences in DNA content of less than 5% can not be detected with present technology on a routine basis and these cases will therefore be classifed as "DNA diploid". Frequently, shoulders of G_1 peaks suggest minimal DNA abnormalities, but these are not considered to be sufficient evidence for DNA aneuploidy according to the recently established guidelines [12].

The correlation between FCM derived DNA stemlines and cytogenetically determined karyotype has recently been investigated [7, 11] and was found to be excellent. The two methods complement each other in the detection of clonal abnormalities in leukemia. FCM is proliferation-independent in its recognition of aneuploid G_1 peaks, but limited in its sensitivity. Cytogenetics is more sensitive in the detection of minor abnormalities but requires mitotic cells for study which are not always available in sufficient quality and quantity in preparations from leukemic samples. Both methods used in combination identify higher numbers of abnormal clones than either method used alone.

Multiple DNA stemlines were rarely seen in our series. In several patients, diploid and aneuploid cells were involved in the leukemic process, and monoclonal antibodies, i.e. against CALLA, were essential in the identification of diploid leukemic cells. One patient presented with four different stemlines and the question arises whether this represents monoclonal or polyclonal disease.

It is generally assumed that one stemline evolves from another by clonal evolution, and all stemlines originate from the same initially transformed cell; we have previously reported one case with "biclonal-biphenotypic" leukemia, i.e. diploid leukemic cells with lymphoblastic features and tetraploid leukemic cells with myeloblastic characteristics [27]. The definitive answer to monoclonal vs. polyclonal origin can only be provided by G-6PD or karyotype analysis. One case with two unrelated leukemic clones has already been described [28]. This patient had two clones: a) 46xx, t(11;14) and b) 46xx, del (6) (q25) when ALL was initially diagnosed. At relapse, only clone (b) was present and the leukemia was classified as AML.

Future studies will clarify the issue of polyclonal leukemia, but the main tool available at the present time, i.e. cytogenetic analysis, is not able to identify very small numbers of cells that may constitute additional clones. Hopefully, cytogenetic analysis of specific cell types analyzed by FCM and then separated by electrostatic cell sorting will allow for a more refined approach to this fundamental problem of leukemogenesis.

When DNA aneuploidy was present, it was a very useful "tumor marker", only marginally influenced by drugs used in the treatment of leukemia, and detected in levels as low as 1–2%. It proved useful for *monitoring:*

1. effects of induction therapy;
2. recovery of regenerating marrow;
3. residual disease after induction therapy, and
4. minimal residual disease during maintenance therapy.

Induction chemotherapy reduced the number of aneuploid cells, and they disappeared in most patients who achieved clinical complete remissions. In some patients, however, aneuploid cells were present when clinical criteria suggested CR and this inevitably led to relapse. The higher number of aneuploid cells as compared to blast cells in pretreatment bone marrows may be the result of the removal of some non-leukemic cells by the Ficoll-Hypaque gradient separation prior to FCM. Alternatively, some of the aneuploid leukemic cells may have the morphological appearance of more differentiated cells and are therefore not counted as leukemic blast cells.

During chemotherapy, the differences became so pronounced that merely technical reasons can be excluded. This may be explained by the relative resistance to chemotherapy of more differentiated leukemic cells and rapid killing of the more immature blast cells. In fact, the observed decrease in proliferating S phase cells and of G_1 cells with high RNA content supports this notion. Alternatively, the concept of *induced differentiation*, well established in leukemic cell lines [29] and also shown to be valid in fresh human myeloid leukemia transferred to an in vitro system, could be invoked to explain the differences: leukemic cells differentiate beyond the blast stage, but maintain their abnormal DNA content. This was previously described for ANLL cells induced in vitro to differentiate into functional macrophages [30]. Cell kill could then be the result of a direct cytotoxic effect of the drugs used, and also of terminal differentiation. A study to elucidate the different mechanisms is currently under way and we expect to get a definitive answer for DNA aneuploid patients using the double staining technique for DNA and differentiation antigens.

In a regenerating bone marrow with immature blast cells, it is often impossible to distinguish normal hematopoietic from leukemic cells. If an aneuploid DNA stemline has been identified prior to treatment, it is then possible to discriminate leukemic from non-leukemic regenerating cells and this is of major clinical importance for therapeutic decisions.

The persistence of cells with abnormal DNA content during and after induction therapy was usually followed by early "relapse". In these cases a true "complete remission" had not been achieved despite fulfillment of all standard criteria. We suggest therefore that the concept and definition of "complete remission" should be modified to incorporate the information available from DNA measurements.

A large number of pediatric patients with DNA aneuploid ALL was monitored for years, during and after maintenance therapy. Ongoing CR was associated with the absence of detectable aneuploid cells, except for two cases (data not shown). These data support the concept of leukemic cell eradication rather than of continuous induced differentiation, within the limits of the sensitivity of FCM and with the assumption that abnormal DNA stemlines of leukemic cells do not revert to diploid stemlines. The low number of relapses at this point precludes definitive evaluation of FCM as a tool to detect relapse earlier than by conventional means, but the detection of CNS relapse by FCM 6 months prior to clinical relapse is encouraging.

Differentiation

We have described a method to simultaneously measure DNA content and differentiation antigens of individual cells by multiparameter FCM and demonstrated its application to human leukemia. It allows us to better classify acute leukemia by use of a panel of monoclonal antibodies, to investigate leukemia cell heterogeneity, and in particular to better define "biphenotypic" leukemia, i.e. the presence of more than one leukemic phenotype, and "biclonal" leukemia (Fig. 4). The use of monoclonal antibodies to identify early and late myeloid antigens will lead to a taxonomic system for ANLL that is based on cell biology rather than on morphological criteria. Of particular interest is the direct correlation of differentiation and cell proliferation [31] and we are presently using this techniques to study the differentiating effects of individual drugs that are known to effect the cell cycle, i.e. ARA-C, and of standard induction regimens.

Proliferation

The proliferative activity of leukemic cells has been studied for many years, initially with autoradiographic methods and more recently by FCM. Several new observations are noteworthy: in pediatric ALL, the proliferation of leukemic cells with abnormal DNA content was significantly higher than of those with normal DNA content, and the RNA content of aneuploid $G_{0/1}$ cells was likewise increased compared to diploid cells [21].

The rate of DNA aneuploidy and proliferation was higher in the FAB L2 subtype, as compared to the L1 phenotype. No difference in S-phase was found between adult ALL and ANLL. Pretreatment and recovery bone marrow S phase, determined in biopsy material which is not contaminated with peripheral blood cells [13], was not predictive of response in adult ANLL patients treated with standard induction protocols (DAT or AAT, Table 2). The nadir bone marrow biopsy S-phase, however, was significantly lower in patients who achieved CR as compared to those who failed to respond, indicative of a more complete eradication of proliferating cells in successfully treated patients. We observed similar changes in patients treated with high-dose ARA-C: decrease of S-phase was associated with clinical response but increased S-phase indicated resistance to ARA-C manifested by cell cycle block or delay rather than a cytotoxic effect [32].

Outlook

FCM of leukemia has significantly increased our understanding of the biology of human acute leukemia and will probably continue to do so in conjunction with other techniques.

Recently, specific chromosomal abnormalities have been identified for a variety of human leukemias, some with a surprising specificity for leukemic subtypes as identified by the French-American-British system [34]. In certain leukemias, breakpoints of the involved chromosomes have also been identified as the sites of oncogens [33]. The human cellular homologue of the Abelson murine leukemia virus is normally located on chromosome 9 and is translocated to chromosome 22q– in the Philadelphia translocation typical for chronic myelogenous leukemia [35]. In the same translocation the sis oncogene, which codes for the platelet derived growth factor (PDGF), is translocated from chromosome 22 to chromosome 9 [36]. In Burkitt's lymphoma a piece of chromosome 8 is moved to another chromosome, usually 14 but sometimes 2 or 22 [34]. The myc oncogene has been localized on the breakpoint of chromosome 8 [37]. The role of these oncogenes and their correlation with cell growth is presently under investigation. We have recently developed an assay to simultaneously measure the expression of the ras-oncogene coded protein p21 in individual cells and their DNA content by multiparameter flow cytometry.

FCM will continue to quantitate biologically important cellular features and probes for specific gene products will enable us to better understand leukemia on a molecular level.

Acknowledgement

We would like to express our gratitude to Ms. Edith Espiritu, Ms. Susan Swartwout, Ms. Jan Bressler, Mr. James Steinmetz and Mr. Gordon Assing for their excellent technical assistance in performing these studies. Dr. Mannoni generously provided the monoclonal antibodies against myeloid cell surface antigens. Drs. Benjamin Koziner and Ayad Al-Katib performed some of the cell surface marker studies. Dr. J. Jett kindly analyzed DNA histograms and Mr. Sal Leto prepared the manuscript.

Drs. Lois Murphy, Norma Wollner, Paul Meyers, Timothy Gee, Sanford Kempin, Roland Mertelsmann, David Strauss and Ellin Berman supplied samples for these studies and Ms. Susan McKenzie did the cytochemical studies. Dr. Bayard D. Clarkson was a continuous source of encouragement, inspiration and constructive criticism.

This work was supported by grants from the National Cancer Institute CA-20194, CA-29564 and CA-05826. Dr. Redner is a fellow of the Leukemia Society of America.

References

1. Andreeff M (1974) Wirkung verschiedener neuer Cytostatica auf die DNA-Reduplikation und die Proliferationskinetik am Modell des Ehrlich-Lettreschen Ascites-Tumors. Verh Dtsch Ges Innere Medizin 80:1678–1680
2. Andreeff M (1975) Impulscytophotometrische DNS-Bestimmung proliferierender Systeme. In: Andreeff M (ed) Impulscytophotometrie. Springer Verlag, Berlin Heidelberg New York, p 73–76
3. Andreeff M (1977) Cell Kinetics of Tumor Growth. Fundamentals–Methods–Experiments (in German), Stuttgart
4. Buechner TH, Barlogie B, Asseburg U, Hiddemann W, Kamanabroo D, Goehde W (1974) Accumulation of S phase cells in the bone marrow of patients with acute leukemia by cytosine arabinoside. Blut 28:299–300
5. Barlogie B, Latreille J, Fu CT, Franco J, Meistrich M, Andreeff M (1980) Characterization of hematologic malignancies by flow cytometry. Blood Cells 6:714–744
6. Hiddemann W, Buechner T, Andreeff M, Woermann B, Melamed MR, Clarkson BD (1982) Cell kinetics in acute leukemia. A critical re-evaluation based on new data. Cancer 50:250–258
7. Barlogie B, Raber MN, Schumann J, Johnson TS, Drewinko B, Swartzendruber DE, Goehde W, Andreeff M, Freireich EJ (1983) Flow cytometry in clinical cancer research. Cancer Res 43:3982–3997
8. Look AT, Melvin SL, Williams DL, Brodeur GM, Dahl CV, Kalwinsky DK, Murphy SB, Mauer AM (1982) Aneuploidy and percentage of S-phase cells determined by flow cytometry correlate with cell phenotype in childhood acute leukemia. Blood 60:(4)959
9. Andreeff M, Darzynkiewicz A, Sharpless TK, Clarkson BD and Melamed MR (1980) Discrimination of human leukemia subtypes by flow cytometric analysis of cellular DNA and RNA. Blood 55:282–293
10. Kruth HS, Braylan RC, Benson NA, Nourse VA (1981) Simultaneous analysis of DNA and cell surface immunoglobulin in human B-cell lymphoma by flow cytometry. Cancer Res 41:4895–4899
11. Andreeff M, Conjalka M, Jhanwar S, Middletown A, Redner A, Assing G, Miller D, Clarkson B, Melamed MR, Chaganti R (1982) Clonal abnormalities in acute lymphoblastic leukemia: Comparison of cytogenetics and flow cytometry. Blood [Suppl 1] 60:120a
12. Hiddemann W, Schumann J, Andreeff M, Barlogie B, Herman C, Leif RC, Mayall BH, Murphy RF, Sandberg AA (1984) Convention on nomenclature for DNA cytometry. Cytometry 5, 445–446
13. Hiddemann W, Buechner T, Andreeff M, Woermann B, Melamed MR, Clarkson BD (1982) Bone marrow biopsy instead of "marrow juice" for cell kinetic analysis. Comparison of bone marrow biopsy and aspiration material. Leuk Res 6:601–612
14. Andreeff M (1975) Teilsynchronisation durch Bleomycin. Untersuchung der DNS-Synthese und Zellzykluskinetik am Ehrlich-Lettre-Ascites-Tumor. In: Wilmanns W (ed) Bleomycin, Illertissen, S 69–78

15. Dolbeare F, Gratzner HG, Pallavicini MG, Gray JW (1983) Flow cytometric measurement of total DNA content and incorporated bromodeoxyuridine. Proc Natl Acad Sci USA 80:5573
16. Andreeff M, Cirrincione C, Assing G, Melamed MR, Clarkson BD (1984) Analysis of prognostic factors including flow cytometry derived parameters in adult acute non-lymphoblastic leukemia (ANLL). Proc Am Soc Clin Oncol 3:758
17. Clarkson BD, Ellis S, Little C, Gee T, Arlin Z, Mertelsmann R, Andreeff M, Kempin S, Koziner B, Chaganti R, Jhanwar S, McKenzie S, Cirrincione C, Gaynor J (1985) Acute lymphoblastic leukemia in adults. Sem Oncol (in press)
18. Mannoni P, Janowska-Wieczorek A, Turner AR, McGann L, Turc JM (1982) Monoclonal antibodies against human granulocytes and myeloid differentiation antigens. Human Immunol 5:309–323
19. Doerner MH, Broxmeyer HE, Silverstone B, Andreeff M (1983) Biosynthesis of ferritin subunits from different cell lines of human promyelocytic leukemia cells, HL-60, and the release of acidic isoferritin-inhibitory activity against normal granulocyte-macrophage progenitor cells. Br J Haematol 55:47–58
20. Mertelsmann R, Moore M, Clarkson B (1982) Leukemia cell phenotype and prognosis: an analysis of 519 adults with acute leukemia. Blood Cells 8:561–583
21. Suarez C, Andreeff M, Miller DR, Steinherz PG, Melamed MR (1985) DNA and RNA determinations in 111 cases of childhood acute lymphoblastic leukemia (ALL) by flow cytometry: Correlation of FAB classification with DNA stemline and proliferation. Br J Haematol (in press)
22. Thongprasert S, Molander D, Andreeff M (1983) Changes in proliferation and RNA content during induction classification of acute non-lymphoblastic leukemia (ANLL) determined by multiparameter flow cytometry. Proc Am Assoc Cancer Res 24:8
23. Redner A, Melamed MR, Steinherz P, Miller D, Murphy L, Andreeff M (1983) Comparison of aneuploid cells in bone marrow aspirate (Asp) and biopsies (Bx) in acute leukemia monitoring. Proc Am Assoc Cancer Res 24:8
24. Redner A, Andreeff M, Miller DR, Steinherz P, Melamed MR (1984) Recognition of CNS leukemia by flow cytometry. Cytometry 5:614–618
25. Andreeff M, Miller D, Steinherz P, Kempin S, Straus D, Clarkson B (1981) Hypodiploid acute lymphoblastic leukemia (ALL): A rare entity detected and monitored by flow cytometry. Proc Am Assoc Cancer Res [Abstract] 22:175
26. Clarkson B, Gee T, Arlin Z, Mertelsmann R, Kempin S, Dinsmore R, O'Reilly R, Andreeff M, Berman E, Higgins C, Little C, Cirrincione C, Ellis S (1984) Current status of treatment of acute leukemia in adults: an overview. In: Büchner Th, Urbanitz D, van de Loo J (eds) Therapie der akuten Leukämien. Springer Verlag, Berlin Heidelberg New York
27. Andreeff M, Gee T, Mertelsmann R, McKenzie S, Steinmetz J, Chaganti R, Koziner B, Clarkson B (1980) Biclonal lymphoblastic and myeloblastic acute leukemia. Proc Am Assoc Cancer Res [Abstract] 21:213
28. Williams DL, Dahl GV, Boroman WP, et al. (1982) Karyotype features of the chromosome ploidy group in childhood acute lymphoblastic leukemia (ALL). Evidence for nonrandom abnormalities relating to prognosis. Blood [Suppl 1] 60:141a
29. Chiao JW, Freitag WB, Steinmetz JC, Andreeff M (1981) Changes of cellular markers during differentiation of HL-60 promyelocytes to macrophages as induced by T lymphocyte conditioned medium. Leuk Res 477–489
30. Chiao JW, Andreeff M, Freitag WB, Arlin Z (1982) Induction of in vitro proliferation and maturation of human aneuploid myelogenous leukemia cells. J Exp Med 54:1397
31. Bloch A (1984) Induced cell differentiation in cancer therapy. Cancer Treat Rep 68:199–205
32. Andreeff M, Kempin S, Arlin Z, Mertelsmann R, Espiritu E, Gee T (1983) High dose cytosine-arabinoside (HDARAC) in acute leukemia (AL). Correlation of clinical response and cell kinetics. Proc Am Assoc Cancer Res 24:166
33. Slamon DJ, de Kernion JB, Verma IM, Cline MJ (1984) Expression of cellular oncogenes in human malignancies. Science 244:256–262
34. The fourth international workshop on chromosomes in leukemia, 1982 (1984) Cancer Genet Cytogenet 11:251–360

35. de Klein A, van Kessel AG, Grosfeld G, Bartram CR, Hagemeijer A, Bottsma D, Spurr NK, Heisterkamp N, Groffen J, Stephenson JR (1982) A cellular oncogene is translocated to the Philadelphia chromosome in chronic myelocytic leukemia. Nature 300:765–767

36. Groffen J, Heisterkamp N, Stephenson JR, van Kessel AG, de Klein A, Grosfeld G, Bootsma D (1983) c-sis is translocated from chromosome 22 to chromosome 9 in chronic myelocytic leukemia. J Exp Med 158:9–15

37. Rowley, JD (1982) Identification of the constant chromosome regions involved in human hematologic malignant disease. Science 216:749–751

38. Andreeff M, Bressler J, Higgins P (1985) Onkogene und Krebs: Übersicht und neue Methode zur Messung der Genexpression in Abhängigkeit vom Zellzyklus. Dtsch med Wschr 110:30–35

Flow Cytometry as a Diagnostic and Prognostic Tool in Cancer Medicine

B. Barlogie

Introduction

Microscopic cell and tissue inspection plays an important role in the diagnostic evaluation of patients and is unsurpassed in cancer diagnosis. Whereas the biochemical examination of body fluids and tissue extracts has long reached an objective and quantitative stage, morphologic cell and tissue investigation remains subject to pathologist expertise and bias. Over the past decade, automated cytology using both image analysis and flow cytometry has been greatly advanced, enabling an objective and quantitative assessment of those cellular features that are crucial to microscopic diagnosis [1–4]. In this review, we discuss the use of flow cytometry as a diagnostic and prognostic tool to supplement morphology for the management of neoplasia. Principally, a flow cytometer consists of a flow chamber, a light source, an array of optical lenses, and electronic signal processing devices. Properties amenable to flow cytometric evaluation include light scatter [5], fluorescence [6] and electric impedance [7]. Light scatter has been found useful both for measurement of cell size (forward angle scatter) and cell structure (right angle scatter), whereas fluorescence intensity provides accurate information on the quantity of cellular constituents as long as specificity and stochiometry of dye binding are assured. These requirements are readily fulfilled in the case of nucleic acid and protein analysis, whereas immunofluorescence measurements are not as clearly defined.

In reviewing the clinical applications of flow cytometry, we will cover the areas of cell and tissue processing; delineate the major cytometric probes currently available for the diagnosis of neoplasia and those pertinent to differentiation and proliferation; and finally assess the value of such cytometric features for specific diagnosis and prediction of clinical outcome.

Cell and Tissue Processing

A major requirement for cell analysis by flow systems is complete monodispersal of tissue, so that a quantitative analysis of cellular properties can be conducted on a cell-by-cell basis. Among the cell dispersal techniques currently in use, both mechanical and enzymatic procedures have extensively been applied with varying success [8]. It is difficult to make a general recommendation other than pointing out

---

Professor of Medicine, M.D. Anderson Hospital and Tumor Institute, Department of Hematology, 6723 Bertner – BX 55 Houston, TX 77030, USA

Tumor Aneuploidy
Büchner et al.
© Springer-Verlag: Berlin Heidelberg 1985

that they need modification according to tissue source and desired cytometric parameters. In the case of nuclear DNA content analysis, rather crude enzymatic treatment with pepsin will still preserve the nuclear membrane while partially destroying the cytoplasm [9]. For cell surface analysis, enzymatic procedures should be avoided, in order to preserve the antigenic constituents that are subject of investigation. Likewise, in the case of hormone receptor analysis, active transport mechanisms affording nuclear translocation of the receptor complex may be altered by enzymatic treatment that affects cytoplasmic function. For cell material already in single cell suspension such as peripheral blood and bone marrow aspirate, red cell removal and syringing generally suffice [10]. In the case of solid tissue, we have been using enzymatic treatment with pepsin at room temperature following mechanical mincing as our preferred method for high resolution DNA content analysis [9, 11]. In the case of concurrent DNA and RNA content analysis, employing acridine orange, mechanical dispersal has generally been the sole dispersal method [12].

Following cell monodispersal, a variety of other steps are required, depending on the cellular constituents to be measured. In the case of DNA content analysis, prior ethanol fixation and RNA digestion are generally employed to achieve quantitative and specific DNA staining by a number of intercalating or base pair specific dyes [10, 13, 14]. In the case of concurrent DNA-RNA analysis, on the other hand, viable cells are made permeable to acridine orange by Triton X treatment [12]. Rather than going into the details of staining protocols, I will refer to specific published procedures in the review of the various cellular markers.

Cytometric Probes

Neoplastic Markers

There is a considerable body of evidence both from single cell cytophotometric and more recently flow cytometric investigation that an abnormal DNA content is a reliable indicator of neoplasia [11, 15]. None of the reactive tissue lesions or benign tumors have yet been demonstrated to display an abnormal DNA content. A word of caution has to be raised with regard to some reports indicating that DNA-related fluorescence is a function of chromatin condensation, sex and age [16, 17]. With the highly DNA-specific dye combination of ethidium bromide (intercalating agent) and mithramycin (G-C-specific agent) employing energy transfer, such dependences were not noted [10]. An important factor in defining abnormalities in DNA content relates to the use of proper biological standards. After appropriate instrument adjustment using fluorescent spheres, we have been using human lymphocytes or granulocytes as diploid reference standard mixed in 2 different ratios with the experimental sample [18]. Under high resolution measurement conditions (co-efficient of variation < 2%), we require the demonstration of 2 distinct subpopulations for the diagnosis of an abnormal DNA content. Other authors have employed non-human reference standards such as chicken erythrocytes with markedly lower DNA content as well as trout erythrocytes characterized by higher DNA content than normal human diploid cells [19]. We believe that the direct admixture of test samples and human reference cells undergoing identical preparative and staining procedures is probably best suited for a conclusive diagnosis of abnormal DNA content. The

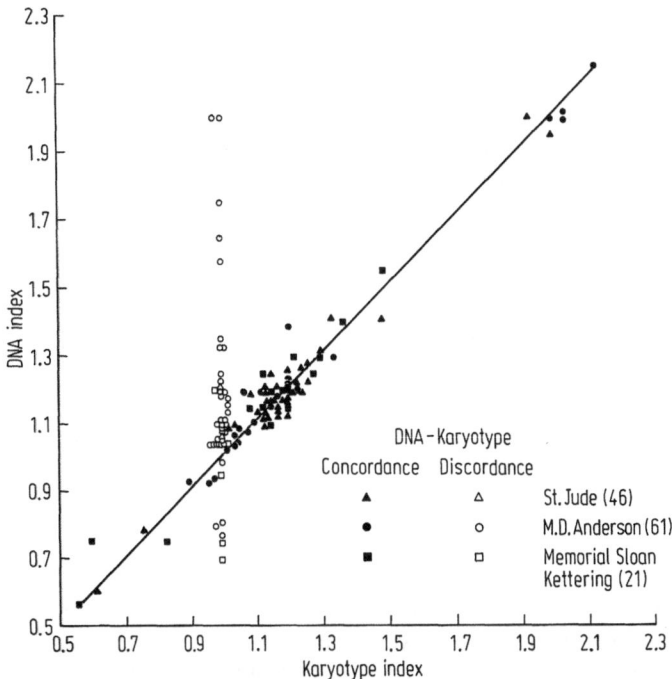

Fig. 1. Relationship between DNA content and karyotype in 128 patients with acute leukemia, expressing an abnormal DNA stemline. The data are expressed as DNA index (ratio of modal G_1 DNA content of tumor cells vs diploid reference standard) and karyotype index (ratio of modal chromosome number and 46). There is a linear correlation for 85 patients between DNA and karyotype index values (●, ▲, ■). A DNA-karyotype discordance with striking abnormalities in DNA content in the presence of a normal diploid or near-diploid karyotype was observed in 43 patients (○, △, □). These two principal subsets of patients with abnormal DNA index were seen in separate studies from 3 different institutions (St. Jude, Thomas A. Look; SKCC, Michael Andreeff; and MDAH, Bart Barlogie). (Reproduced with permission from Barlogie B, et al.: Cancer Res 43:3982, 1983)

chance of detection of abnormal DNA stemlines increases with measurement resolution, degree of DNA content abnormality, and the proportion of DNA-abnormal cells. A complicating circumstance is the overlap between abnormal G_1 cells with DNA excess and normal diploid cells undergoing DNA synthesis. Thus, the detection of DNA-abnormal stemlines is also influenced by the degree of proliferation of normal tissue cells present in a tumor lesion.

We have introduced the term DNA index as a measure of the degree of DNA content abnormality, expressing simply the ratio of DNA fluorescence of abnormal to normal $G_{1/0}$ cells [18]. We were able to demonstrate that the proportion of cells with abnormal DNA index values in leukemia and in myeloma generally correlates well with the proportion of microscopically identified leukemic blast and myeloma plasma cells, respectively [19, 20]. In the leukemias, we also showed that the degree of DNA content abnormalities parallels that of chromosome number for the majority of patients (Fig. 1) [15, 18]. However, there was a separate cohort with a numerically normal karyotype that exhibited disproportionally high degrees of DNA content ab-

Table 1. Aneuploidy in human malignancies

Diagnosis	No. of patients	Percent aneuploid
Leukemia	595	22
Lymphoma	360	53
Myeloma	177	76
Colon cancer	135	62
Breast cancer	385	79
Lung cancer	353	85
Prostate cancer	147	62
Bladder cancer	459	82
Testis cancer	74	93
Melanoma	643	76
Sarcoma	41	98
All solid tumors	3559	75
All malignancies	4691	67

normality [15]. This discrepancy between discretely abnormal DNA content of G_1 cells and numerically normal metaphase karyotype could, in some instances, be related to a growth disadvantage of leukemic cells, so that only residual normal diploid hemopoietic cells would be available for cytogenetic analysis in mitosis. Using the technique of premature chromosome condensation [21], we could indeed demonstrate numeric chromosomal abnormalities in G_1 cells of a similar degree noted by DNA content analysis [18]. This observation stresses the advantage of DNA flow cytometry for the detection of numeric chromosomal abnormalities, independent of proliferative characteristics, as the measurement is performed on $G_{1/0}$ cells comprising generally more than 75% of the cell cycle distribution.

Table 1 summarizes the incidence of DNA-abnormal stemlines in specific disease categories. Overall, a 75% incidence of aneuploidy was noted in more than 3,500 solid tissue neoplasms investigated, contrasting with a substantially lower frequency of 22% in almost 600 cases of acute leukemia and an intermediate frequency of approximately 50% in the malignant lymphomas. In comparison to near-diploid DNA content abnormalities in hematologic malignancies, solid tissue neoplasms display the entire range from near-haploid to hyperoctaploid DNA stemlines (Fig. 2). However, no diagnosis-specific DNA index pattern has yet emerged. In the few studies reporting serial ploidy measurements at diagnosis and at relapse, that is often at the time of drug-resistance, a change in DNA index is the exception.[22, 23]. Likewise, differences in DNA stemlines by different tissue sites in disseminated malignancies are relatively rare and have been reported in patients with malignant melanoma, who demonstrate a comparatively high incidence of multi-clonal abnormalities in primary lesions, whereas metastatic sites usually display unimodal abnormalities. Overall, as in spontaneous dog tumors, multimodal DNA stemline abnormalities occur with a frequency of less than 10% [11, 24]. Thus, compared to the well-known dispersion in chromosome number, DNA stemlines are discrete and unimodal in the majority of almost 5,000 cases surveyed [15, 25].

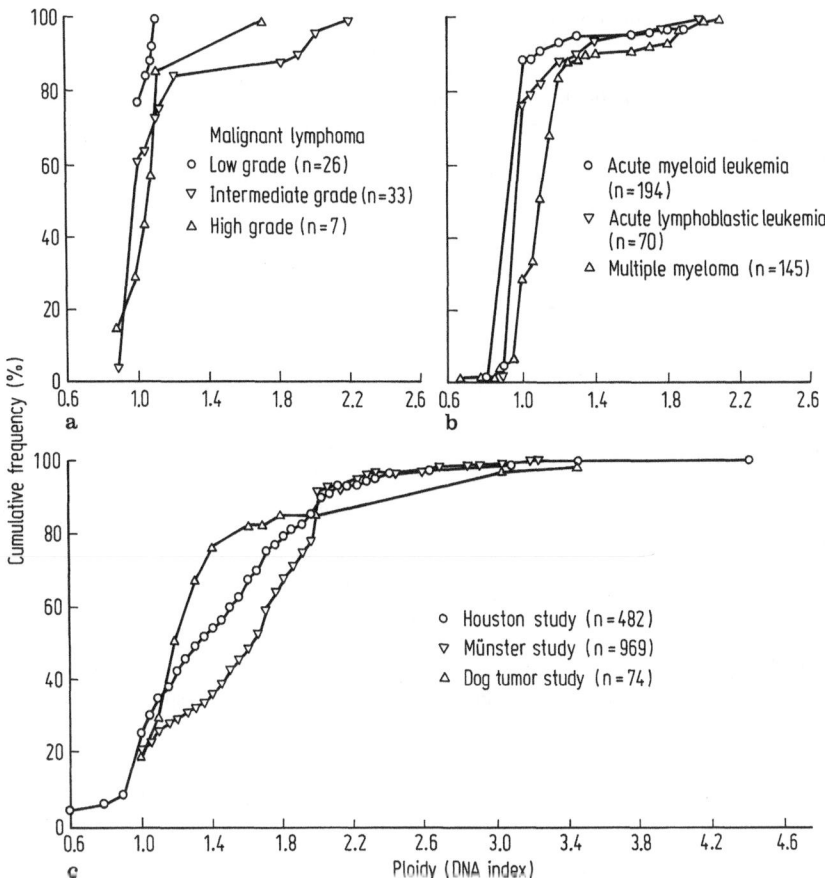

Fig. 2. Cumulative frequency distribution of ploidy levels (DNA index; see text) in non-Hodgkin's lymphoma (A), in acute leukemia and myeloma (B), and in solid tumors (C). Note the prevalence of diploid and low-degree hyperdiploid DNA index values in lymphoma, leukemia, and myeloma, contrasting with an even distribution of DNA index values across the entire spectrum from hypodiploidy to hyperoctaploidy in solid tumors. There is a remarkable similarity in DNA index frequency profiles in solid-tissue neoplasms among 2 large human studies performed in Houston and Münster as well as between human and canine tumors (15)

Alternative Neoplastic Markers

For DNA-diploid disease, alternative cellular features have been sought that are indicative of neoplasia. The most widely studied probe is the nucleolar antigen, as initially described by Busch and his associates [26, 27]. Using antiserum raised in rabbits against HeLa cell nucleoli, these investigators demonstrated by indirect immunofluorescence microscopy the almost universal expression of a nucleolar antigen in a wide variety of human neoplasms, including leukemia, lymphoma and the entire spectrum of solid tumors [28]. There have been occasional observations of nucleolar antigen positive normal appearing cells in leukemic patients in remission, where the issue arises of residual leukemic blasts vs terminally differentiated leu-

111

Table 2. DS-RNA expression in various malignancies

Diagnosis	No. of patients	% DS-RNA excess Median	Range	% patients with DS-RNA excess > 30
AML				
Diagnosis	22	47	20–92	86
Relapse	5	48	34–92	100
ALL				
Diagnosis	4	37	31–54	100
Relapse	8	42	24–92	75
CML				
Benign phase	2	38	2–65	50
Blast crisis	2	67	41–92	100
MULTIPLE MYELOMA Plasma Cell %				
< 50	23	13	0–56	26
> 50	15	44	8–83	60
LYMPHOMA				
Low grade	6	22	11–85	17
Intermediate-High Grade	10	79	27–90	90
SOLID TUMORS				
MISCELLANEOUS	5	56	29–78	80
NORMAL BONE MARROW	23	15	0–33	9
NORMAL LYMPH NODE	2	24	22–26	0

kemic cells [29]. Efforts to apply this indirect immunofluorescence technology to flow cytometry have been met with limited success. There was an overlap in fluorescence intensity between normal diploid and aneuploid tumor cells, possibly because the nucleolar antigen-related fluorescence is not confined to the nucleus but may also arise in the cytoplasm, which is not readily appreciated microscopically but is readily detected by the highly sensitive flow cytometric method [15].

We have also evaluated the use of the intercalating agent propidium iodide to measure double-stranded RNA in whole cells after DNA digestion [30, 31, 15]. We noted significantly higher levels of double-stranded RNA content in 67 patients with a variety of different tumors when compared to normal lymphocytes and marrow controls (Table 2). In contrast, however, to DNA analysis, ds-RNA content shows considerable dispersion in a given tumor; and in approximately 37% of cases, values within the normal tissue range were identified. On microscopic examination, the double-stranded RNA-related propidium iodide fluorescence often localizes in the nucleolus, but a considerable proportion of patients also revealed in addition cytoplasmic fluorescence. Further studies are needed to probe the nature of propidium iodide fluorescence after DNA removal in more detail. We do know already, however, that the degree of double stranded RNA-related fluorescence is only minimally influenced by cytokinetic properties [32].

Differentiation Markers

The rapid expansion of cell surface marker research and its application to normal and malignant hemopoiesis and lymphopoiesis have recently been reviewed [33]. Thus, particularly in the non-Hodgkin lymphomas, efforts have been undertaken to pinpoint the T or B cell nature and the stage of differentiation of malignant lymphomas in comparison to the normal lymphoid maturation pathway. The stage of malignant transformation may be early in lymphopoiesis, whereas the phenotypic presentation of a lymphoma depends on its differentiation potential. Such studies have been conducted both on tissue sections using immunoperoxidase technology and more recently have also employed flow cytometry. Approximately 75 to 80% of lymphomas have been recognized to display B cell properties, 15% carry T cell antigens, and the remaining proportion has an undifferentiated phenotype. Using an ever increasing panel of monoclonal antibodies, substages of both T and B cell lymphomas have been recognized, the importance of which in terms of histopathologic correlation and prognosis remains to be determined. Similar considerations apply to the leukemias, where the recent availability of monoclonal antibodies against myeloid surface antigens will lead to extensive investigation of immunologically defined phenotypes in relationship to specific morphologic and karyotypic patterns.

Cellular RNA Content

Cellular RNA can be measured by way of fluorescence either with acridine orange or pyronine Y [12, 34]. Using the former fluorochrome, both DNA and RNA can be measured simultaneously, due to the metachromatic properties of acridine orange: green fluorescence is observed upon interaction of acridine orange with nucleic acids by intercalation, whereas red fluorescence results from dye stacking to single stranded nucleic acid or yet unknown mechanisms [35]. Following selective denaturation of double-stranded RNA which may be induced at a certain acridine orange concentration in the presence of EDTA and/or citrate, a stochiometric relationship between the intensity of green fluorescence and DNA content per cell as well as the intensity of red fluorescence and RNA content have been demonstrated [36, 37].

When applied to the study of human acute leukemias, myeloid phenotypes were found to have significantly higher levels of RNA compared to cases of acute lymphoblastic leukemia (Fig. 3) [38]. Using a RNA index as the ratio of mean RNA content of G_1 cells vs the median RNA content of lymphocytes [38], remarkable heterogeneity was observed in the acute leukemias and in CML (Fig. 4). Median values at diagnosis for AML (2.0) and ALL (1.5) were significantly different, and both were higher than the median value for normal marrow (0.9) [39, 40]. At diagnosis, 83% of 128 patients with AML had RNA index values > 1.5, whereas 60% of 51 patients with ALL had values ≤ 1.5. RNA index values in smouldering or oligoleukemia were similar to those in CML benign phase at diagnosis. The RNA index of myeloid blast crisis was similar to that of AML. Chronic lymphocytic and hairy cell leukemia had RNA index values indistinguishable from those of normal marrow. However, while there was considerable dispersion in RNA content of normal marrow, CLL and HCL typically had a homogeneous RNA content.

Acridine orange-derived measurement of RNA content of normal human marrow using cell separation techniques revealed an increasing RNA content in the order of

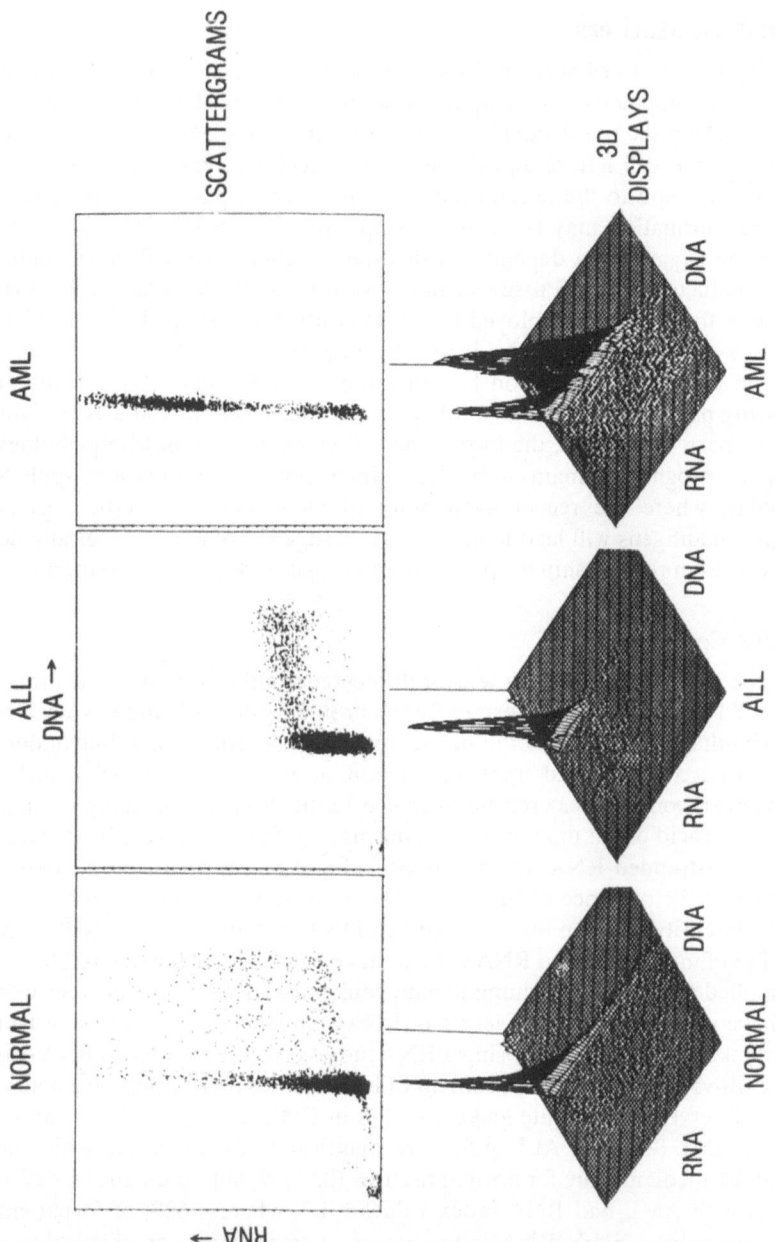

Fig. 3. DNA-RNA frequency distribution of marrow biopsy material from a normal volunteer and a patient each with acute lymphoblastic leukemia (ALL) and with acute myeloblastic leukemia (AML). Depicted are both scattergrams and 3-dimensional displays. Normal marrows are characterized by a predominance of cells with low RNA content with a typical trailing towards higher RNA content. ALL cells display a homogenous low RNA content profile, and proliferation is often limited to the cell subpopulation with the highest RNA content. AML is characterized by a dominant cell subpopulation with markedly elevated RNA content, clearly distinguishable from ALL and normal marrow

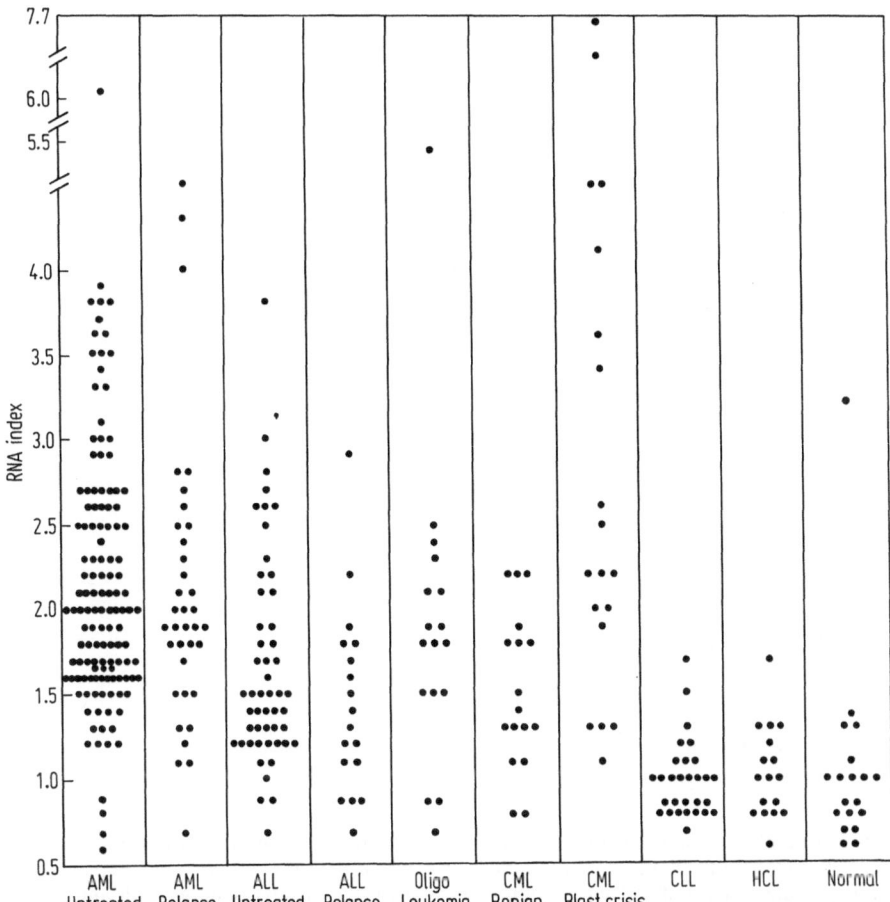

Fig. 4. RNA index in human leukemia. There is considerable heterogeneity in RNA values for most leukemic subtypes, except in chronic lymphocytic leukemia (CLL) and hairy cell leukemia (HCL). Compared to normal marrow, significantly higher RNA index values are noted for all types of acute leukemia, oligoleukemia and chronic myelogenous leukemia (CML). At diagnosis, acute myeloblastic leukemia (AML) displays significantly higher RNA index values compared to acute lymphoblastic leukemia (ALL). An additional distinguishing feature between different leukemia subtypes is provided by cytokinetics (see Fig. 8)

granulocyte/lymphocytes to erythroid and myeloid precursors, whereas plasma cells were characterized by the highest RNA content [39]. In addition to these cell lineage-related differences, RNA content has also been noted to relate to cell size and to cell cycle kinetics [41]. Specifically, RNA content increases from early to later stages in G_1 and permits the distinction of G_{1A} and G_{1B} compartments [41]. However, the individual contributions of cell type, cell size and proliferative characteristics are difficult to assess when dealing with a heterogeneous cell population as is present in normal or leukemic bone marrows.

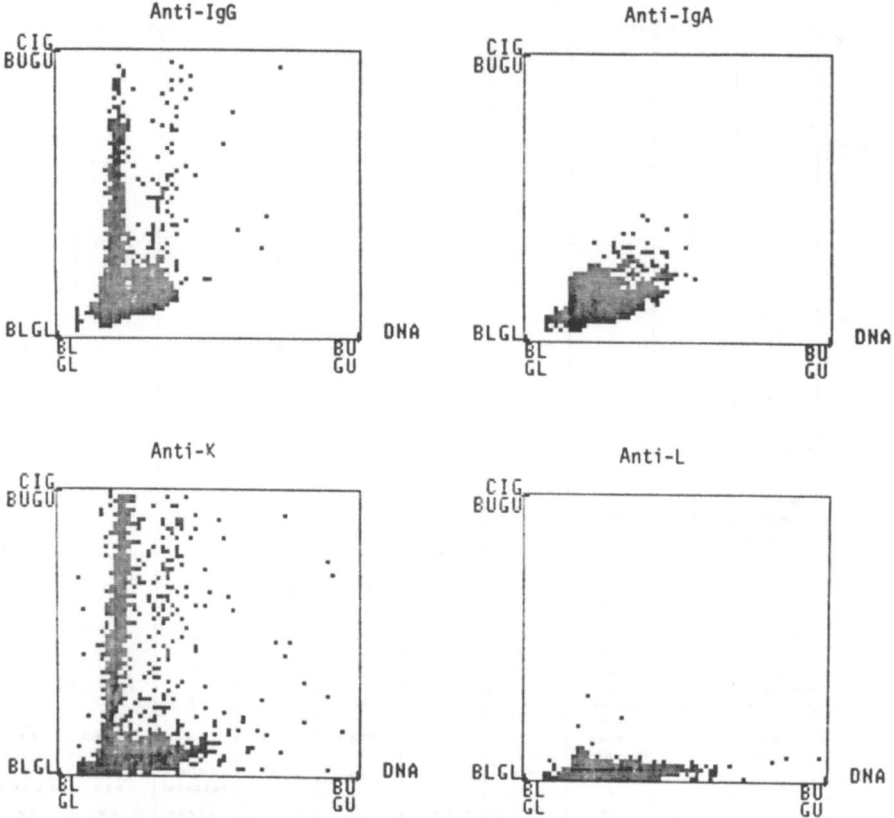

Fig. 5. Cytoplasmic immunoglobulin analysis in a patient with IgG kappa myeloma. Marrow cells were ethanol fixed and doubly stained for DNA (propidium iodide) and cytoplasmic immunoglobulin using indirect immunofluorescence (FITC). Positive staining with anti-gamma heavy chain and anti-kappa light chain antisera strikingly contrasts with background fluorescence noted in the case of incubation with anti-alpha and anti-lambda reagents

RNA content has also been found to be useful in the characterization of non-Hodgkin lymphomas, where a progressive increase in RNA index was noted with increasing histopathologic grade [42]. Finally, the highest RNA content is found in marrow plasma cells from patients with myeloma [43].

Cytoplasmic Immunoglobulin

Cytoplasmic immunoglobulin is a distinctive feature of pre-B and plasma cells and can be readily measured by the use of either direct or indirect immunofluorescence with appropriate anti-heavy and -light chain antibodies (Fig. 5) [44].

Hormone Receptors

A classic example of tumor differentiation is the expression of hormone receptors or, in a broader sense, of receptors for physiologic signals including growth factors. The prime example in this area is human breast cancer, where biologic and prognostic

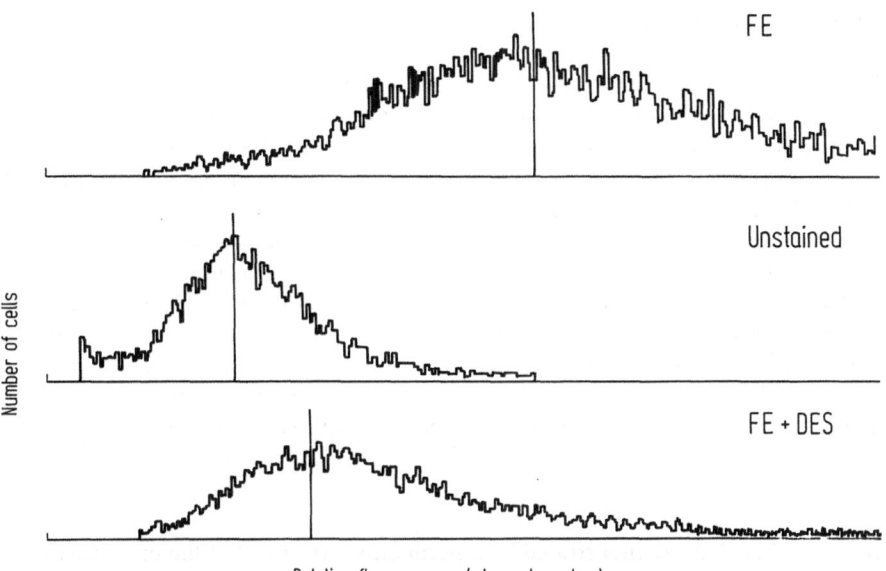

FE

Unstained

FE + DES

Number of cells

Relative fluorescence (channel number)

Fig. 6. Flow cytometric analysis of estrogen receptor content in human breast cancer cells (MCF-7). Using a FITC-conjugated estradiol compound (FE), markedly higher fluorescence intensity is observed than in unstained cells (autofluorescence). FE fluorescence was readily suppressed in case of co-incubation with diethylstilbestrol (DES) to almost autofluorescence levels (non-specific staining)

implications of hormone receptor expression have been amply demonstrated. Our laboratory has recently developed a cytometric assay for estrogen receptor (ER) analysis, using fluorescently labelled estradiol (Fig. 6) [45]. We were able to demonstrate that this compound conforms to the requirements of a receptor ligand. Concordant analysis of cellular DNA content and ER expression by flow cytometry will enable us to investigate the heterogeneity of ER expression related to tumor cells (almost 90% of human breast cancers have an abnormal DNA content) and to cell proliferation. Because of the relatively small amount of tissue required for such analysis, serial studies of ER expression should be possible in patients even with small biopsiable lesions, in order to assess the potential modulation of receptor activity under the influence of both hormonal and cytotoxic therapy.

Other Markers

A number of additional phenotypic parameters of potential biologic and clinical interest can be assessed by flow cytometry, including CEA, alpha-fetoprotein and various enzymes (using fluorogenic substrates) [46].

Cytokinetic Markers

Extensive experimental studies *in vitro* and *in vivo* have demonstrated the major cytokinetic determinants for tumor cell kill, including cycle stage distribution, cycle traverse rate and growth fraction [47].

Cycle Stage Distribution

Besides being a useful marker for the detection of numeric chromosomal abnormalities, cellular DNA content also discriminates the various cell cycle stages and has extensively been used for this purpose both in experimental systems and in human disease [48]. Serial DNA cytometric studies allow the recognition of interference with cycle progression by antitumor agents [49], an approach that has been applied in clinical trials of tumor cell synchronization. It should be pointed out that the proportion of cells in any particular phase of the cell cycle is a direct reflection of the relative phase durations, so that for example the observation of a high S phase fraction does not necessarily imply a high proliferative activity, but can also be the result of ineffective DNA synthesis, as is the case in pernicious anemia. A number of algorithms have been developed for analyzing cell cycle distributions [50–52], and difficulties still exist in case of mixtures of cells with different DNA stemlines.

Cell Cycle Progression

We and others have demonstrated in experimental systems that the presence of cells in a particular cell cycle stage *per se* does not confer a high degree of sensitivity to cytotoxic agents exerting their greatest lethal activity in the corresponding cell cycle stage [53]. In the case of the typical S phase-specific antimetabolite cytosine arabinoside and even for the cycle-active anthracycline antibiotic adriamycin, suspension of cycle progression by a protein synthesis inhibiting agent (anguidine) protected against tumoricidal effects [53]. There are now several means of cell cycle traverse rate analysis using flow cytometry, including modification of fluorescence of a number of DNA-specific dyes following incorporation of the DNA precursor 5'-bromodeoxyuridine [54]. Alternatively, BUdR antibody immunofluorescence technology has been developed to determine the intensity of DNA synthesis across S phase [55].

Growth Fraction

Under standard DNA staining conditions, cycling G_1 cells cannot be distinguished from non-cycling G_0 cells, which constitute a major sanctuary for tumor sterilization by most chemotherapeutic agents and ionizing radiation. The Sloan Kettering group has developed a cytochemical method that allows distinction of non-cycling from cycling cells on the basis of differences in chromatin condensation with resulting differences in the cellular sensitivity to heat or acid denaturation [56]. Employing acridine orange after RNA removal, the highly condensed G_0 and mitotic cells proved to be more sensitive to denaturation than G_1 and S phase cells so that, for a given level of total fluorescence, the proportion of red fluorescence, representing single-stranded DNA, was higher for cells in G_0 and mitosis (Fig. 7). This technique can also identify non-cycling cells with an S phase DNA content, as has been previously suspected on the basis of combined single cell DNA cytophometric and autoradiographic studies [57]. Unfortunately, when applied to the investigation of clinical material, we have noted considerable differences in denaturability as a function of cell type, which may interfere with the ability to distinguish cytokinetic differences [15].

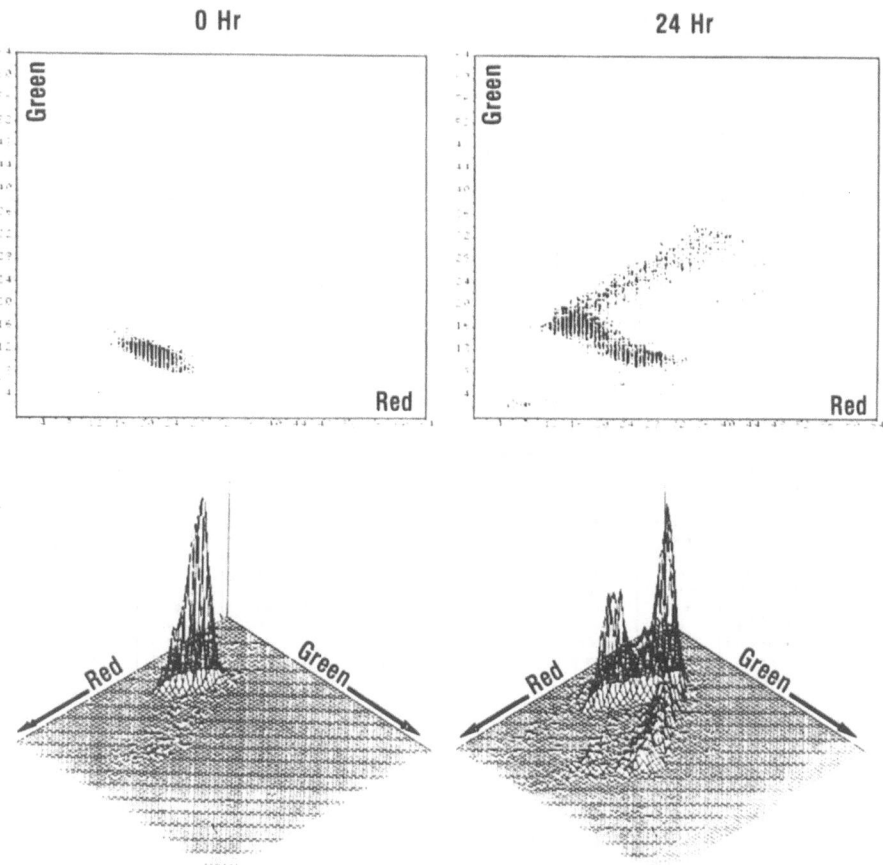

Fig. 7. Effect of *in vitro* culture on red-green fluorescence distribution pattern. While the direct sample reveals a single-cell population, 24 hr culture in Roswell Park Memorial Institute Medium 1640 results in emergence of G_1 cells with higher green and lower red fluorescence intensity compared to the original sample. Also note the population of S-phase cells originating from the high-green-, low-red-fluorescence G_1 population

Cellular Pharmacology and Biochemistry

A number of antitumor agents are fluorescing and are therefore amenable to flow cytometric quantitation. In the case of adriamycin, however, fluorescence quenching has been noted upon its interaction with nuclear DNA [58, 59]. Alternatively, we have recently demonstrated the feasibility of measuring amsidine content by way of competition with the G-C specific DNA fluorochrome Hoechst 33342 [60].

Most antitumor agents act on DNA as a target, and some drugs cause DNA-DNA or DNA-protein cross-links. In preliminary studies with cisplatinum, we could demonstrate that, under condition of increasing cross-linking, the denaturability by acid treatment was progressively reduced, which could be readily probed with acridine orange [40].

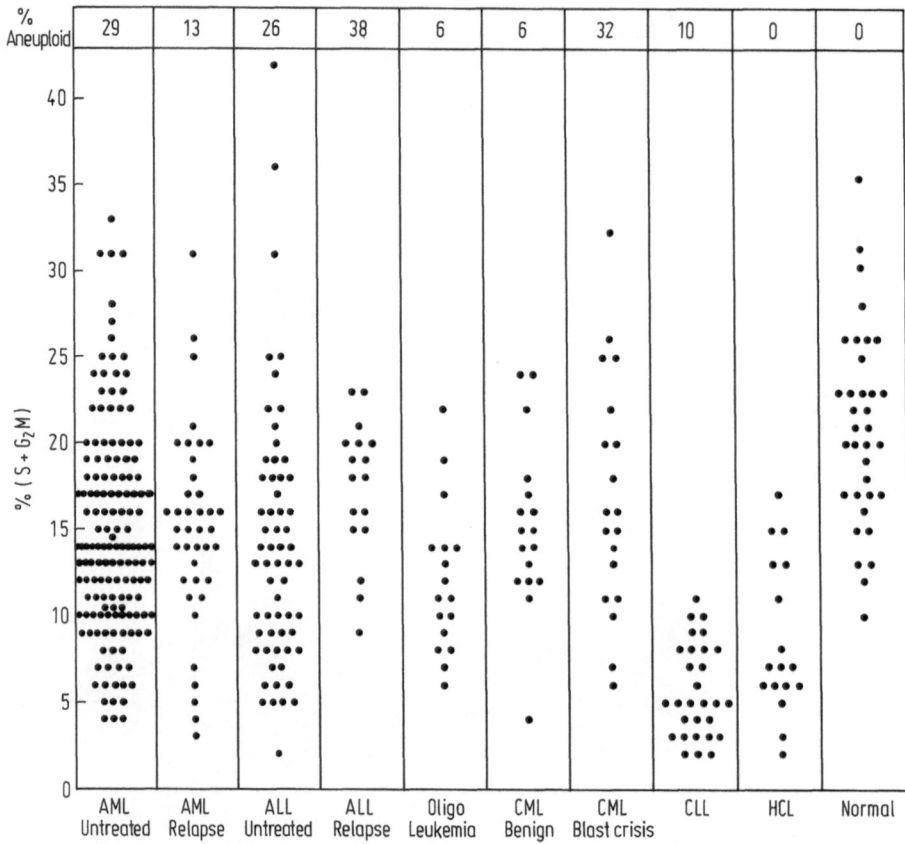

Fig. 8. Cytokinetic properties in bone marrow biopsy material of human leukemias. Considerable heterogeneity in the proportion of cells in $(S + G_2M)$ phase is noted in all categories including normal marrow, except chronic lymphocytic leukemia (CLL). Compared to normal marrow, acute leukemias have significantly lower $(S + G_2M)$ values. Within the acute leukemia group, relapsed AML and particularly ALL are characterized by significantly higher $(S + G_2M)$ values than observed at diagnosis

Flow Cytometry in the Study of Specific Neoplasms

Leukemias

Abnormal DNA stemlines are expressed in approximately 35% of acute lymphoblastic and in 20% of acute myeloid leukemias, but rarely in CML and almost never in CLL or hairy cell leukemia (Fig. 8). DNA-derived S phase proportions typically show low values in CLL and hairy cell leukemia: 97% of patients have values $\leqq 10\%$, contrasting with significantly higher values and a greater degree of heterogeneity in the remaining leukemic presentations and in normal marrow. In comparison to normal bone marrow, 71% of patients with AML and ALL at diagnosis have S phase fractions $\leqq 15\%$, whereas 85% of normal individuals display values $> 15\%$

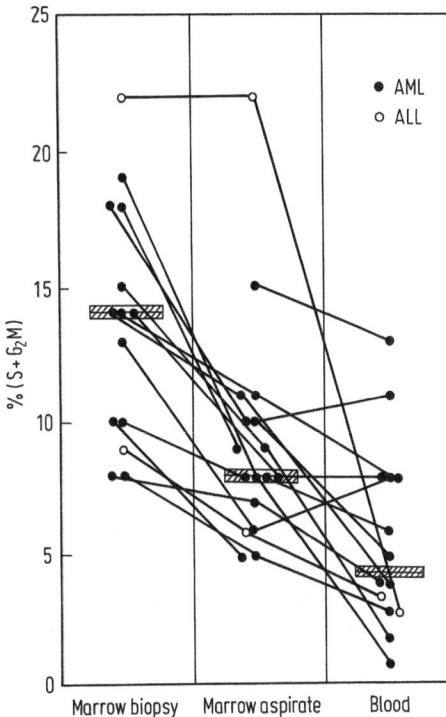

Fig. 9. Cytokinetics in bone marrow and peripheral blood in acute leukemia. Marrow biopsy material is characterized by markedly higher values of $(S+G_2M)\%$ compared to marrow aspirates, which in turn display a higher proliferative activity than leukemic cells in the peripheral blood

($p < 0.01$). It is important in this context to point out that marrow biopsy material consistently contains a larger proportion of cells in S phase than corresponding marrow aspirate due to peripheral blood contamination (Fig. 9).

The heterogeneity in cell cycle distribution characteristics among the different forms of leukemia precludes its use for differential diagnosis. In conjunction with RNA content analysis, however, discrimination between AML and ALL is afforded in almost 80% of patients (Figs. 4 and 8). CLL and hairy cell leukemia were characterized by homogeneous and low RNA content which, together with low S phase values, was discriminatory for most patients studied (Figs. 4 and 8). We have recently added to the RNA index parameter of G_1 cells (see above) also information on the RNA content characteristics of the remaining cell cycle compartments and noted different patterns of increase in RNA content with progression through S phase particularly between AML/ALL and normal marrow. Thus, overt acute leukemia had a significantly steeper increase in RNA during S and G_2 phase than was found in morphologically normal marrow from normal volunteers and from patients with leukemia in remission (Fig. 10). Ultimately, we are aiming to develop a computer-based system that will permit automatic DNA-RNA pattern recognition and assignment to the most likely diagnostic category.

Prognostically, the value of DNA- and RNA-derived parameters in acute leukemia have not yet been fully established. We have noted for remission induction a favorable impact of increasing S% in patients under age 50, whereas the reverse was

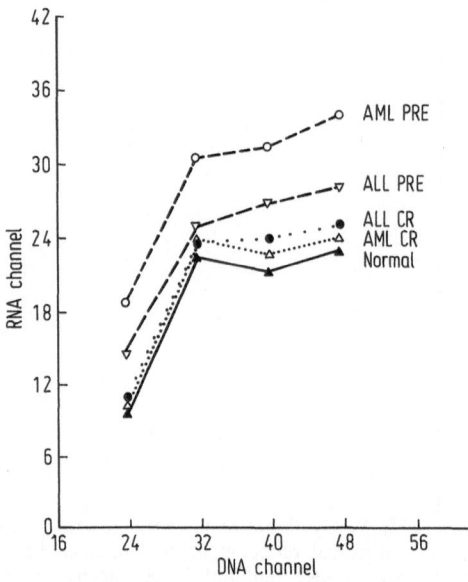

Fig. 10. RNA characteristics as a function of cell cycle stage for cohorts of normal volunteers and AML and ALL both at diagnosis and in remission. Morphologically normal marrow from normal volunteers and remission patients display a fairly constant RNA content during S phase. In contrast, patients with acute leukemia of both myeloblastic and lymphoblastic types demonstrate an increase in RNA content during S phase. Moreover, AML revealed significantly higher RNA values than ALL

true for the older patient population (Fig. 11) [60a]. The basis for this diverse influence of cell cycle kinetics on short term prognosis is not fully understood but seems to be related to differences in cytogenetic characteristics. During remission induction, responding patients tend to display a faster decrease in RNA index and S% followed by an earlier rebound in S% after hypoplasia to values typical for normal marrow [40]. Remission duration in patients with AML is longer when the pretreatment proportion of cells in S phase is low (< 12%), reflecting the importance of regrowth kinetics for the duration of disease control. The pretreatment RNA index seems to affect remission duration in patients with ALL, where higher values more similar to AML are a favorable disease characteristic. We also studied DNA and RNA features at the onset of complete remission and noticed that higher proportions of cells in S were a favorable parameter for remission duration, possibly reflecting a lower leukemic burden and hence a lower level of leukemia inhibitory activity (LIA) targeted towards residual normal hemopoiesis [29]. Finally, double-stranded RNA excess beyond the normal marrow range at the onset of complete remission also adversely affected remission duration (Fig. 12) [29].

Lymphoma

Using appropriate immunofluorescence technology, the B or T cell nature of malignant lymphomas can be readily determined. The availability of monoclonal anti-

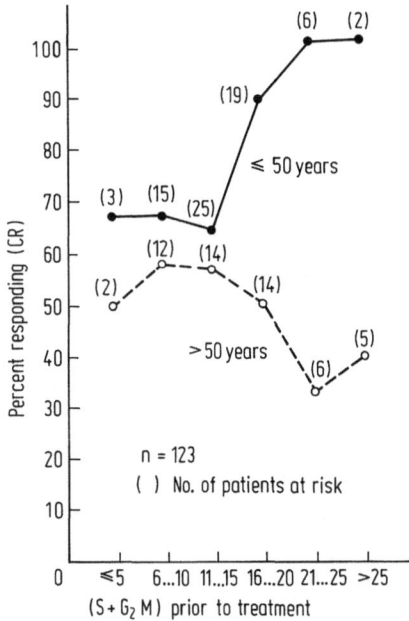

Fig. 11. Complete remission rate in adult AML as a function of cytokinetics and age. While the proportion of cells in $(S + G_2M)$ at diagnosis did not impact on remission induction in the overall study population, high $(S + G_2M)\%$ favors higher remission rates in young patients, but was detrimental for the older age group

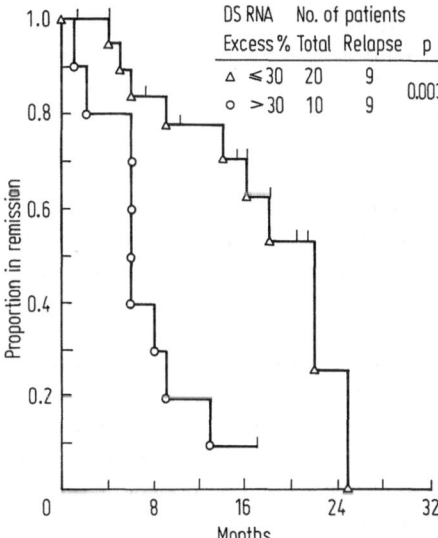

Fig. 12. Impact of double-stranded RNA excess (see text) of marrow cells at the onset of complete remission on remission duration. Patients exhibiting values outside of normal marrow range ($> 30\%$) have a significantly shorter remission duration than those with values $\leqq 30\%$

bodies reacting with lymphoid cells in defined stages of the B or T cell differentiation pathway permits further subclassification, the clinical relevance of which remains to be determined. A complicating circumstance in the flow cytometric evaluation of lymph node material from malignant lymphoma patients relates to the admixture of normal lymphoid cells particularly in the nodular histologies. It will be highly desirable to have available, for such studies of the interaction of malignant

Table 3. Flow cytometry in lymphoma

Histology grade	No. of patients	% aneu-ploid	P	% (S+G₂M)		P	RNA index		P
				Mean	SD		Mean	SD	
Low	67	17		4.7	3.1		0.97	0.43	
			<0.004			<0.001			<0.001
Intermediate	79	52		12.5	8.3		1.65	1.17	
High	17	47		26.2	10.7		1.77	0.65	

and normal lymphoid cells, an unequivocal neoplastic marker such as the nucleolar antigen or an abnormal DNA stemline.

The frequency of abnormal DNA stemlines in malignant lymphoma increased with the histologic grade: aneuploidy was present in 17% of low grade and in approximately one half of aggressive lymphomas (Table 3). The same DNA parameter providing histologic grade-associated ploidy information also affords cytokinetic characterization, which permits an even better discrimination of histologic grades: 90% of low grade histologies have S phase values below 5%, constrasting with 85% of intermediate and high grade lymphomas expressing S phase proportions above 5%. Employing the metachromatic acridine orange dye, the simultaneously acquired RNA content analysis also reveals progressively higher values with increasing histologic grade (Table 3). Thus, by providing joint information on ploidy, cell cycle stage distribution and RNA content, a single flow cytometric analysis of acridine orange-stained cells permits excellent discrimination among the different histopathologic grades as defined in the International Working Formulation [42, 61]. Within the intermediate grade category, diffuse poorly differentiated lymphocytic and large cell histologies according to Rappaport can be distinguished on the basis of homogeneous low RNA content features in the former and more dispersed and higher RNA content as well as a higher frequency of abnormal DNA stemlines in the latter group, both displaying similar S% values (Fig. 13).

Not examined in detail, but of considerable biological and clinical importance is the analysis of involved lymphnodes from multiple sites as well as other involved organ sites. In this fashion, the nature of composite and discordant lymphomas can be further evaluated. Likewise, serial evaluation over time would be of great interest to unravel the cellular changes associated with histologic transformation of low grade lymphomas, especially with regard to genotypic evolution manifested in a change of DNA stemlines. At this moment, there does not appear to be a relationship between any of the nucleic acid-defined parameters and tumor burden defined by either clinical or pathologic stage.

Prognostically, we and others are just now beginning to identify early trends. The presence of DNA-aneuploidy in large cell lymphoma adversely affects the likelihood of remission induction with salvage therapy of previously treated patients (36 vs 68% in 28 patients each with aneuploid or diploid DNA content; p < 0.05). In the same histologic group, survival of patients with DNA-aneuploid disease was significantly shorter (18 months) than that of individuals with normal diploid DNA con-

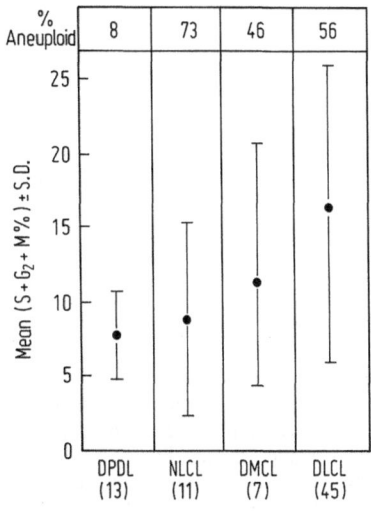

Fig. 13. Cytometric characteristics in intermediate grade lymphoma. Diffuse poorly differentiated lymphocytic lymphoma (DPDL) is characterized by a significantly lower incidence of aneuploidy compared to the large cell and mixed cell histologies. In addition, there is a progressive increase in the average proportion of cells in $(S+G_2M)\%$ with increasing degree of aggressiveness within the intermediate grade histologies

tent (36+ months). It is conceivable that, ultimately, nucleic acid information along with surface antigen characterization will adequately reflect the tremendous clinical heterogeneity among the malignant lymphomas, so that treatment intensity can be properly gauged.

Myeloma

The microscopic evaluation of marrow plasmacytosis has not been widely used for staging of patients with myeloma or their management because of presumably patchy marrow involvement by tumor and difficulties in unequivocal identification of myeloma plasma cells using routine Giemsa staining.

Employing initially DNA content analysis and later on concurrent analysis of DNA and RNA content, we have demonstrated occurrence of DNA-abnormal stemlines in approximately 80% of 250 patients studied, and a discretely elevated RNA content was noted in 90% of 150 patients, thereby objectively identifying also DNA-diploid tumor cells [22, 43]. In comparing marrow biopsy and aspirate material, the latter tissue source surprisingly showed higher proportions of tumor cells when compared to marrow biopsies in the 10% of patients with discrepant findings [9]. This observation contrasts with our experience in leukemia, where biopsy material gives more representative information with regard to cytokinetic parameters. From investigations of multiple sites and correlation studies with clinically derived tumor mass stage, it has been our experience that patchy marrow involvement by tumor is the exception and pertains to < 10% of all patients. Indeed, using the clinical staging system first introduced by Durie and Salmon, a correlation was observed with marrow tumor infiltrate as determined by flow cytometry (Fig. 14) [9]. In comparison to microscopic assessment of marrow plasmacytosis, the cytometric determination of marrow tumor infiltration showed less overlap between the different clinical stages of tumor burden.

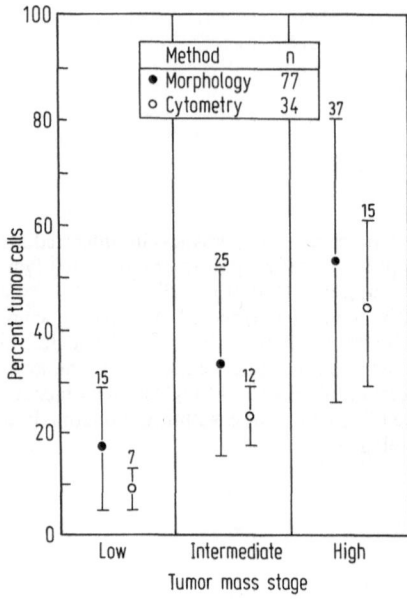

Fig. 14. In patients with multiple myeloma studied at diagnosis, the degree of marrow infiltration increases significantly with advancing tumor mass stage. Using DNA-RNA cytometry, the separation among the different tumor mass groups is superior to that provided by morphologic analysis

Table 4. DNA-RNA cytometry in extramedullary plasmacytoma

Bone marrow	No. of patients	Extramedullary plasmacytoma		Bone marrow discordance	
		Percent aneuploid	Percent RNA high	DNA	RNA
Uninvolved	7	86	100	NA	
Involved	6	100	100	20	67

Prognostically, RNA content of myeloma tumor cells was identified as a novel cellular feature to affect the likelihood of initial treatment response, thus providing the first prognostic test for remission induction (Fig. 15) [9]. It was the degree of marrow tumor infiltrate, on the other hand, that determined the length of survival as expected, in view of an excellent correlation between marrow infiltrate and clinical tumor mass stage (Fig. 16) [9].

In contrast to some other investigators' results, we did not find the degree of ploidy abnormality in myeloma to strongly affect short or long term prognosis [43, 63]. Likewise, we generally noted a return of the same genotypic abnormality upon development of drug resistance at the time of clinical relapse [22].

We had the opportunity to examine the genotypic and phenotypic features of myeloma tumor cells in marrow in relationship to those of accompanying plasmacytomas of bone or in soft tissue (Table 4). There were both concordant and discordant DNA stemlines, which have to be investigated in the context of clonal evo-

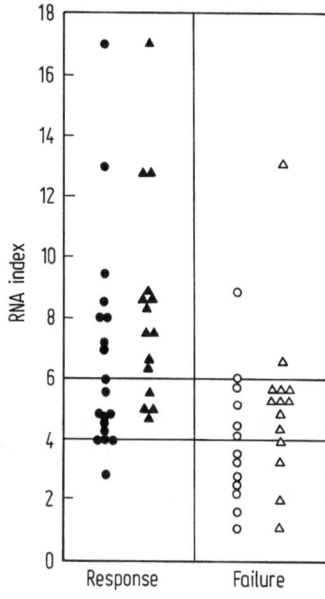

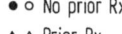

Fig. 15. The RNA index of myeloma plasma cells in the marrow is an important prognostic factor for response both in previously untreated and in treated patients. Thus, there is a progressive decrease in response rate as RNA index values decrease from > 6 to the range of 4 to 6 to values ≦4

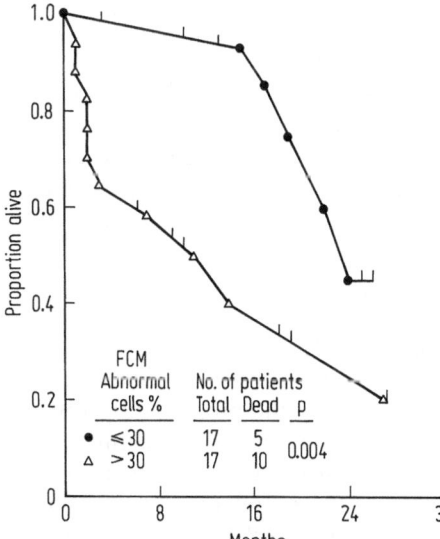

Fig. 16. Survival in previously untreated myeloma is significantly affected by the portion of cytometrically abnormal cells in the marrow: patients with an infiltrate ≦30% survive significantly longer than individuals with values > 30%

lution. In a small series of 30 successive patients, we examined the peripheral blood for DNA-abnormal stemlines and found such abnormalities in 20% of patients, including 3 individuals studied at diagnosis. Thus, there is evidence of circulating aneuploid tumor cells early during the disease course in myeloma. In the absence of morphologically recognizable plasma cells, this finding raises the question of circulating neoplastic precursor cells deserving further investigative attention.

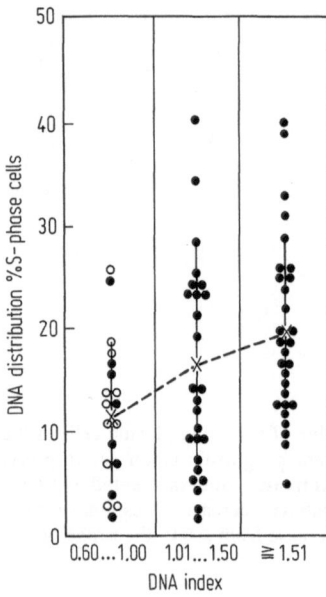

Fig. 17. Relationship between proliferation (%S) and ploidy (DNA index) in primary breast cancer. There is a significant increase in the median proportion of cells in S phase with increasing DNA index. Patients with DNA-diploid characteristics are symbolized by open circles

Breast Cancer

Human breast cancer has been studied extensively from the point of view of prognostic factors. Disease-free and overall survival following primary surgery have been linked to size of the primary tumor, number of involved regional lymph nodes, degree of hormone receptor expression and proliferative activity. The latter two tumor cellular parameters seem to have an independent prognostic impact. Once metastatic disease develops, the response to hormonal therapy is greatly influenced by the degree of estrogen and progesteron receptor expression and is virtually absent in patients with receptor-negative disease.

We have studied the triangle relationship between ploidy, proliferation and estrogen receptor content as a differentiation marker in both patients with primary tumors and in individuals with metastatic disease (64). The incidence of DNA-abnormal stemlines was identical (85%) in both primary and metastatic disease. In primary breast cancer, we noted a significant increase in the proportion of cells in S phase with increasing degree of ploidy abnormality (Fig. 17). Furthermore, the S phase compartment size decreased with increasing degree of ER positivity (Fig. 18). Of great interest to us was the observation that the inverse relationship between proliferative activity and estrogen receptor expression only pertained to premenopausal patients. At this moment, we do not have information available on the joint impact of the 3 different variables along with standard clinical parameters on disease-free and overall survival in a study population of more than 200 patients, as the follow-up is still too short.

A research direction of great interest will be the analysis of hormone receptor content on a per cell basis, using fluorescently labelled hormones or monoclonal antibodies directed against the receptor. Our own work has been with an FITC-con-

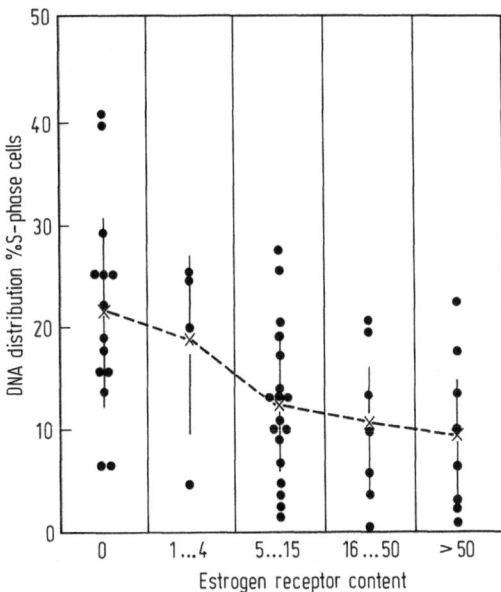

Fig. 18. Relationship between proliferative activity and estrogen receptor (ER) content (fmol/mg). With increasing receptor expression, a decline in the proportion of cells in S phase is noted, which is particularly obvious for pre-menopausal patients (see text)

jugated estradiol compound kindly provided by Barrows [65]. We were able to demonstrate the feasibility of this probe to determine estrogen receptor content and are now attempting to relate ER expression to DNA content [45]. This will afford ER analysis on a per tumor cell basis in 85% of human breast cancer specimens as well as examination of cell cycle stage dependent fluctuations.

Lung Cancer

An area of great interest is the objective discrimination between the various histologic types of lung cancer, so that the suspected difference in their histogenesis and the question of mixed cell histology versus induced differentiation of small cell tumors can be further addressed. In this regard, a number of cellular markers have been identified to distinguish oat and non-oat cell lung cancer [66, 67].

Several investigators have evaluated the role of nucleic acid content analysis to probe for differences in ploidy and proliferative activity [68–70]. Among 193 tissue samples from patients with lung cancer, abnormal DNA stemlines were noted in 70% of 87 cases of small cell and in 90% of 106 patients with non-oat cell lung cancer, with bi-clonal frequencies of 6 and 12%, respectively. There was a similar distribution in DNA-ploidy levels for oat cell and non-oat cell lung cancer. When analyzed by primary tissue source, a significantly higher aneuploidy rate was noted in surgical and bronchoscopy specimens as compared to bronchial washings, and sputum contained the lowest incidence of DNA-abnormal cells (Table 5). In a comparison between microscopic and cytometric examinations of 59 oat cell carcinomas, concordance for primary lung cancer specimens was noted in 77%, and discordance concerned 20% of patients displaying FCM-abnormal/cytology-normal findings and 3% with the contrary FCM-normal/cytology-abnormal characteristics. As illustrated

Table 5. DNA-aneuploidy in oat cell lung cancer as a function of sample procurement

Tissue source	n	% aneu-ploid
Sputum	14	50
Bronchoscopy		
Washing	14	57
Brushing	11	73
Surgical biopsy	10	60
Biopsy	10	70
Total	59	61

Table 6. Ploidy and percent S in lung cancer at diagnosis

Histology	n	DNA index			P value
		≤ 1.0	1.01–1.45	> 1.46	
Oat cell	87	$8.4^a \pm 4.3$	13.1 ± 5.9	20.6 ± 6.6	0.01
Non-oat cell	100	7.5 ± 4.3	16.7 ± 8.9	23.8 ± 7.0	0.01

[a] Median %S$\pm$SD

in Table 6, the proportion of cells in S phase increases with increasing DNA-ploidy values in both oat and non-oat cell histologies. Prognostically, however, response to chemotherapy in oat cell carcinoma or survival corrected for standard clinical variables such as stage, performance status and weight loss have not yet been noted to be influenced by ploidy or cytokinetic parameters.

Other Solid Tumors

A number of investigators have thoroughly evaluated bladder and prostate cancer as well as adenocarcinoma of the colon [71]. In the area of bladder cancer, poorly differentiated tumors have been noted to display both higher aneuploidy rates [72] and higher proliferative activity [73]. The Sloan-Kettering group has demonstrated the value of DNA-RNA flow cytometry for monitoring of *carcinoma in situ* lesions [74]. The Karolinska group, on the other hand, has provided convincing evidence that the degree of DNA excess in prostate carcinoma is directly related to histologic grade and survival [75].

Summary and Conclusion

In conclusion, flow cytometry has gradually emerged as a powerful tool for quantitative description of such relevant biological cell features as ploidy, proliferation and cell differentiation, and a number of important clinical applications in the areas of cancer diagnosis, differential diagnosis and prognosis have already been identified. At present, the single most reliable marker of neoplasia is an abnormal DNA stemline, and it remains to be seen whether the nucleolar antigen and/or double-stranded RNA content can be firmly established as alternate markers in situations of DNA-diploid disease. Specific tissue diagnosis can only be accomplished in those malignancies that express a unique cellular feature such as the T and B-cell antigens in the lymphomas that are readily assessable by appropriate immunologic techniques. Total cellular RNA content is useful for the distinction of myeloblastic from lymphoblastic leukemias and for the characterization of myeloma plasma cells in the bone marrow. Examination of hormone receptor expression in relationship to DNA-abnormal tumor cells and to cell proliferation represents a particularly fruitful area of studying phenotypic heterogeneity in breast cancer and other hormone-dependent tumors. Prognostically, with few exceptions, high degree DNA excess stemlines and high proportions of cells in S phase seem to represent adverse prognostic factors, although such findings often depend on the uniformity and intensity of treatment employed. For example, the previously poor risk group of patients with aggressive lymphoma currently represents the major subgroup with a realistic prospect for cure. As large enough data bases are being established for many of the major disease categories, a number of tumor cell parameters most likely will be identified to be associated with clinical outcome, independent of the classical measures of tumor burden and host performance. This will then facilitate clinical trial design, aimed at gauging treatment intensity according to a patient's individual prognosis. It is conceivable that flow cytometry eventually will provide a fingerprint information that may supersede the importance of histogenetically determined tumor diagnosis according to which treatment is currently being administered.

Acknowledgement

The authors wish to thank Mattie J. Thomas for her invaluable service in helping to prepare this manuscript.

Supported in part by grants CA28771 and CA28153 from the National Institutes of Health, National Cancer Institute, Bethesda, Md. 20205.

References

1. Wied GL, Bahr GF, Bartels PH (1970) Automatic analysis of cell images by TICAS. In: Wied GL, Bahr GF (eds) Automated Cell Identification and Cell Sorting, Academic Press, Inc, New York, pp 195–360
2. Kamentski LA, Melamed MR, Derman H (1965) Spectrophotometer: new instrument for ultra rapid cell analysis. Science 150:630–631
3. van Dilla MA, Trujillo TT, Mullaney PF, Coulter JR (1969) Cell microfluorometry: a method for rapid fluorescence measurement. Science 163:1213–1214
4. Göhde W, Dittrich W (1971) Impulsfluorometrie – Ein neuartiges Durchflußverfahren zur ultraschnellen Mengenbestimmung von Zellinhaltsstoffen. Acta Histochem [Suppl] 10: 429–437
5. Salzman GC, Hiebert RD, Jett JH, Bartholdi M (1980) High-speed single particle sizing by light scattering in a flow system. SPIE 220:23–27
6. Crissman HA, Steinkamp JA (1973) Rapid simultaneous measurement of DNA, protein, and cell volume in single cells from large mammalian cell populations. J Cell Biol 59:766–771

7. Hoffman RA, Johnson TS, Britt WB (1981) Flow cytometric electronic direct current volume and radiofrequency impedence measurements of single cells and particles. Cytometry 1:377–384
8. Crissman HA, Mullaney PF, Steinkamp JA (1975) Methods and application of flow systems for analysis and sorting of mammalian cells. Methods Cell Biol 9:179–2462
9. Barlogie B, Johnston DA, Smallwood L, Schumann J, Drewinko B (1978) Determination of ploidy and proliferative characteristics of human solid tumors by pulse cytophotometry. Cancer Res 38:3333–3339
10. Barlogie B, Spitzer G, Hart JS, Johnston DA, Büchner T, Schumann J, Drewinko B (1976) DNA-histogram analysis of human hemopoietic cells. Blood 48:245–258
11. Barlogie B, Drewinko B, Schumann J, Göhde W, Dosik G, Johnston DA, Freireich EJ (1980) Cellular DNA content as a marker of neoplasia in man. Am J Med 69:195–203
12. Traganos F, Darzynkiewicz Z, Sharpless T, Melamed MR (1977) Simultaneous staining of ribonucleic and deoxyribonucleic acid on unfixed cells using acridine orange in a flow cytofluorometric system. J Histochem Cytochem 25:46–56
13. Göhde W, Schumann J, Zante J (1978) The use of DAPI in pulse cytophotometry. In: Lutz D (ed) Third International Symposium Pulse Cytophotometry. European Press, Ghent, Belgium, pp 229–238
14. Krishan A (1975) Rapid flow cytofluorometric analysis of mammalian cell cycle by propidium iodide staining. J Cell Biol 66:188–193
15. Barlogie B, Raber MN, Schumann J, Johnson TS, Drewinko B, Swartzendruber D, Göhde W, Andreeff M, Freireich EJ (1983) Perspectives in Cancer Research: Flow Cytometry in Clinical Cancer Research. Cancer Res 43:3982–3997
16. Vindelov LL, Christensen IJ, Jensen G (1983) High resolution ploidy determination by flow cytometric DNA analysis. Results obtained by a set of methods for sample storage, staining and internal standardization. Cytometry 3:332–339
17. Staiano-Coico L, Darzynkiewicz Z, Melamed MR, Weksler M (1982) Changes in DNA content of human blood mononuclear cells with senescence. Cytometry 3:79–83
18. Barlogie B, Hittelman W, Spitzer G, Hart J, Trujillo J, Smallwood L, Drewinko B (1977) Correlation of DNA distribution abnormalities with cytogenetic findings in human adult leukemia and lymphoma. Cancer Res 37:4400–4407
19. Vindelov LL, Christensen IJ, Jensen G (1980) High resolution ploidy determination by flow cytometric DNA analysis with two internal standards. Basic Appl Histochem 24:271
20. Latreille J, Barlogie B, Dosik G, Johnston DA, Drewinko B, Alexanian R (1980) Cellular DNA content as a marker of human multiple myeloma. Blood 55:403–408
21. Johnson R, Rao PM (1970) Mammalian cell fusion. II. Induction of premature chromosome condensation in interphase nuclei. Nature 226:712–722
22. Barlogie B, Latreille J, Swartzendruber DE, Smallwood L, Maddox AM, Raber MN, Drewinko B, Alexanian R (1982) Quantitative cytometry in myeloma research. In: Schmidt WR (ed) Clinics in Hematology. W. B. Saunders Co., Philadelphia pp 19–45
23. Barlogie B, Johnston DA, Smallwood L, Raber MN, Maddox AM, Latreille J, Swartzendruber DE, Drewinko B (1982) Prognostic implications of ploidy and proliferative activity in human solid tumors. Cancer Genetic Cytogenet 6:17–28
24. Johnson TS, Raju MR, Giltman RK, Gillete EL (1981) Ploidy and DNA distribution analysis of spontaneous dog tumors by flow cytometry. Cancer Res 41:3005–3009
25. Kraemer PM, Petersen DF, van Dilla MA (1971) DNA constancy in heteroploidy and the stem line therapy of tumors. Science 174:714–717
26. Busch H, Gyorkey F, Busch RK, Davis FM, Yorke P, Smetana AK (1979) A nucleolar antigen found in a broad range of tumor specimens. Cancer Res 39:3024–3030
27. Davis FM, Gyorkey F, Busch RK, Busch H (1979) Nucleolar antigen found in several human tumors but not in non-tumor tissues studied. Proc Natl Acad Sci USA 76:892–896
28. Smetana H, Busch RK, Hermansky F, Busch H (1981) Nucleolar immunofluorescence in bone marrow specimens of human hematologic malignancies. Blut 42:79–86
29. Barlogie B, Hittelman W, Davis FM, Kantarjian H (1983) Nucleic acid cytometry, interphase chromosome and nucleolar antigen in the detection of residual leukemia in morphology. Proc. International Symposium on Detection and Treatment of Minimal Residual Disease in Acute Leukemia. Rotterdam, Holland, October 12–14

30. Frankfurt O (1980) Flow cytometric analysis of double-stranded RNA content distributions. J Histochem Cytochem 28:663–669
31. Kantarjian H, Barlogie B, Stroehlein J (1983) Preferential expression of double-stranded (DS)-RNA in tumor vs normal cells. Abstract Proc. American Society of Clinical Oncology. San Diego, Ca., May
32. Kantarjian H, Barlogie B, Pershouse M, Swartzendruber DE, Keating MJ, McCredie KB, Freireich EJ (1985) Preferential expression of double-stranded (DS)-RNA in tumor vs normal cells: biologic and clinical implications. Blood (in press)
33. Foon KA, Schroff RW, Gale RP (1982) Surface markers on leukemia and lymphoma cells: Recent advances. Blood 60:1–9
34. Shapiro HM (1981) Flow cytometric estimation of DNA and RNA content in intact cells stained with Hoechst 33342 and Pyronine Y. Cytometry 2:143–150
35. Kapuscinski J, Darzynkiewicz Z, Melamed MR (1982) Luminescence of solid complexes of acridine orange with RNA. Cytometry 2:201–210
36. Coulsen PB, Bishop AP, Lenarduzzi R (1977) Quantitation of cellular deoxyribonucleic acid by flow cytometry. J Histochem Cytochem 25:1147–1153
37. Bauer KD, Dethlefsen LA (1980) Total cellular RNA content: Correlation between flow cytometry and ultraviolet spectroscopy. J Histochem Cytochem 28:493–499
38. Andreeff M, Darzynkiewicz Z, Sharpless T, Clarkson B, Melamed MR (1980) Discrimination of human leukemia subtypes by flow cytometric analysis of cellular DNA and RNA. Blood 55:282–293
39. Barlogie B, Latreille J, Freireich EJ, Fu CT, Mellard D, Meistrich M, Andreeff M (1980) Characterization of hematologic malignancies by flow cytometry. Blood Cells 6:619–744
40. Barlogie B, Maddox AM, Johnston D, Raber MN, Drewinko B, Keating MJ, Freireich EJ (1983) Quantitative cytology in leukemia research. Blood Cells 9:35–55
41. Darzynkiewicz Z, Traganos F, Melamed MR (1980) New cell cycle compartments identified by multi-parameter flow cytometry. Cytometry 1:98–108
42. Srigley J, Barlogie B, Butler JJ, Osborne B, Blick M, Johnston D, Kantarjian H, Reuber J, Batsakis J, Freireich EJ (1985) Heterogeneity of non-Hodgkins lymphoma probes by nucleic acid cytometry. Blood 65:1090–1096
43. Latreille J, Barlogie B, Johnston DA, Drewinko B, Alexanian R (1982) Ploidy and proliferative characteristics in monoclonal gammopathies. Blood 59:43–51
44. Barlogie B, Alexanian R, Pershouse M, Smallwood L, Smith L (1985) Cytoplasmic immunoglobulin content in multiple myeloma. JCI (in press)
45. Van NT, Raber MN, Barrows G, Barlogie B (1984) Estrogen receptor analysis by flow cytometry. Science 224:876–879
46. Dolbeare FA, Smith RE (1979) Flow cytoenzymology: rapid enzyme analysis of single cells. In: Melamed MR, Mullaney PF, Mendelsohn ML (eds) Flow Cytometry and Sorting, New York, John Wiley and Sons, Inc. pp 317–334
47. Barlogie B, Drewinko B, Raber MN, Swartzendruber DE (1982) Cell kinetics in clinical oncology. In: Nicolini C (ed) Cell Growth, Plenum Publishing Corp, New York, pp 773–778
48. Crissman HA, Tobey RA (1974) Cell cycle analysis in 20 minutes. Science 184:1297–1298
49. Barlogie B, Drewinko B (1978) Cell cycle stage dependent induction of G_2 phase arrest by different antitumor agents. Eur J Cancer 14:741–745
50. Dean PN, Jett JH (1974) Mathematical analysis of DNA distributions derived from flow microfluorometry. J Cell Biol 60:523–527
51. Johnston DA, White RA, Barlogie B (1978) Automatic processing and interpretation of DNA distributions: Comparison of several techniques. Comp Biomed Res 11:393–404
52. Baisch H, Beck H-P, Christensen IJ, Hartman NR, Fried J, Dean PM, Gray JW, Jetts JH, Johnston DA, White RA, Nicolini C, Zeitz S, Watson JV (1982) A comparison of mathematical methods for the analysis of DNA histograms obtained by flow cytometry. Cell Tissue Kinet 15:235–249
53. Teodori L, Barlogie B, Drewinko B, Swartzendruber DE, Mauro F (1981) Reduction of Ara-C and adriamycin cytotoxicity following cell cycle arrest by anguidine. Cancer Res 41:1263–1270

54. Böhmer RM (1979) Flow cytometric cell cycle analysis using the quenching of 333258 Hoechst fluorescence by bromodeoxyuridine incorporation. Cell Tissue Kinet 12:101–110
55. Gratzner HG (1982) Monoclonal antibody to 5-bromo- and 5-iodo-deoxyuridine: A new reagent for detection of DNA replication. Science 218:474–475
56. Darzynkiewicz Z, Traganos F, Andreeff M, Sharpless T, Melamed M (1979) Different sensitivity of chromatin to acid denaturation in quiescent and cycling cells revealed by flow cytometry. J Histochem Cytochem 27:478–485
57. Ernst P, Faille A, Killman SA (1973) Perturbation of cell cycle of human leukemic myeloblasts *in vivo* by cytosine arabinoside. Scand J Haematol 10:209–218
58. Krishan A, Ganapathi RN, Israel M (1978) Effect of adriamycin and analogs on the nuclear fluorescence of propidium iodide-stained cells. Cancer Res 38:3656–3662
59. Nooter K, van den Engh G, Sonneveld P (1983) Quantitative flow cytometric determination of anthracycline content of rat bone marrow cells. Cancer Res 43:5126–5130
60. Andersson B, Barlogie B, Van N, McCredie K, Beran M (1985) Analysis of m-AMSA content by DNA-fluorochrome competition. Eur J Cancer (in press)
60a. Kantarjian H, Barlogie B, Keating M, Hall R, Smith T, McCredie KB, Freireich EJ (1985) Pretreatment cytokinetics in acute myelogenous leukemia: age-related prognostic implications. JCI (in press)
61. Diamond LW, Nathwani BN, Rappaport H (1982) Flow cytometry in the diagnosis and classification of malignant lymphoma and leukemia. Cancer 50:1122–1135
62. Barlogie B, Alexanian R, Gehan EA, Smallwood L, Smith T, Drewinko B (1983) Marrow cytometry and prognosis in myeloma. J Clin Invest 72:853–861
63. Bunn PA, Keasnow S, Schlam M, Schechter G (1983) Flow cytometric analysis of DNA content of bone marrow cells in patients with plasma cell myeloma: clinical implications. Blood 59:528–535
64. Raber MN, Barlogie B, Latreille J, Bedrossian C, Fritsche H, Blumenschein G (1982) Ploidy, proliferative activity and estrogen receptor content in human breast cancer. Cytometry 3:36–41
65. Barrows GH, Stroupe S, Gray LA (1978) In vitro uptake and nuclear endometrium. Am J Clin Pathol 70:330–331
66. Doyle A, Cuttitta F (1983) Small cell lung cancer (SCLC) and non-small lung cancer (NSCLC) lines differ markedly in the expression of HLA-frame and B-2 microglobulin (B_2-m) antigen. Proc. American Society of Clinical Oncology, Abstract, March
67. Mulshine J, Cuttitta F, Little C, Fargion S, Fedorko J, Bibro D, Carney M, Matthews A, Gazdar A, Minna J (1983) Murine monoclonal antibodies which selectively bind non-small cell lung cancer (Non-SCLC). Proc. American Society of Clinical Oncology, Abstract, March
68. Johnson TS, Barlogie B, Valdivieso M, Bedrossian C (1982) Ploidy and proliferative activity in human lung cancer. In: Cytometry in the Clinical Laboratory. Eng. Foundation Conference, Santa Barbara, Ca.
69. Vindelov LL, Hansen HH, Christensen IJ, Spang-Thomsen M, Hirsch FR, Hansen M, Nissen MI (1980) Clonal heterogeneity of small-cell anaplastic carcinoma of the lung demonstrated by flow-cytometric DNA analysis. Cancer Res 40:4295–4300
70. Bunn P, Schlam M, Gazdar A (1980) Comparison of cytology and DNA content analysis by flow cytometry (FCM) in specimens from lung cancer patients. Proc Am Assoc Cancer Res 21:160
71. Wolley RC, Schreiber K, Koss LG, Karas M, Sherman A (1982) DNA distribution in human colon carcinoma and its relationship to clinical behavior. J Natl Cancer Inst 69:15–22
72. Tribukait B, Esposti PL (1978) Quantitative flow-microfluorometric analysis of the DNA in cells from neoplasms of the urinary bladder: correlation of aneuploidy with histological grading and cytological findings. Urol Res 6:201–205
73. Tribukait B, Gustafson H, Esposti P (1979) Ploidy and proliferation in human bladder tumors as measured by flow cytofluorometric DNA-analysis and its relations to histopathology and cytology. Cancer 43:1742–1751
74. Klein FA, Herr HW, Whitmore WF Jr, Sogani PC, Melamed MR (1982) An evaluation of automated flow cytometry (FCM) in detection of carcinoma *in situ* of the urinary bladder. Cancer 50:1003–1008
75. Tribukait B (1983) Personal Communication

Subject Index

140